ROCK ENGINEERING SYSTEMS
Theory and Practice

ELLIS HORWOOD SERIES IN CIVIL ENGINEERING

Series Editors

Structures: Professor H.R. EVANS, Department of Civil Engineering, University College, Cardiff

Hydraulic Engineering and Hydrology: Dr R.H.J. SELLIN, Department of Civil Engineering, University of Bristol

Geotechnics: Professor D. MUIR WOOD, Cormack Professor of Civil Engineering, University of Glasgow

Addis, W.	**STRUCTURAL ENGINEERING: The Nature of Theory and Design**
Allwood, R	**TECHNIQUES AND APPLICATIONS OF EXPERT SYSTEMS IN THE CONSTRUCTION INDUSTRY**
Bhatt, P.	**PROGRAMMING THE MATRIX ANALYSIS OF SKELETAL STRUCTURES**
Blockley, D.I.	**THE NATURE OF STRUCTURAL DESIGN AND SAFETY**
Britto, A.M. & Gunn, M.J.	**CRITICAL STATE SOIL MECHANICS VIA FINITE ELEMENTS**
Bljuger, F.	**DESIGN OF PRECAST CONCRETE STRUCTURES**
Carmichael, D.G.	**STRUCTURAL MODELLING AND OPTIMIZATION**
Carmichael, D.G.	**CONSTRUCTION ENGINEERING NETWORKS**
Carmichael, D.G.	**ENGINEERING QUEUES IN CONSTRUCTION AND MINING**
Cyras, A.A.	**MATHEMATICAL MODELS FOR THE ANALYSIS AND OPTIMISATION OF ELASTOPLASTIC SYSTEMS**
Edwards, A.D. & Baker, G.	**PRESTRESSED CONCRETE**
Fox, J.	**TRANSIENT FLOW IN PIPES, OPEN CHANNELS AND SEWERS**
Francis, A.J.	**INTRODUCING STRUCTURES: Civil & Structural Engineering, Building & Architecture**
Graves-Smith, T.R.	**LINEAR ANALYSIS OF FRAMEWORKS**
Gronow, J.R., Schofield, A.N. & Jain, R.K.	**LAND DISPOSAL OF HAZARDOUS WASTE**
Hendry, A.W., Sinha, B.A. & Davies, S.R.	**LOADBEARING BRICKWORK DESIGN, 2nd Edition**
Heyman, J.	**THE MASONRY ARCH**
Holmes, M. & Martin, L.H.	**ANALYSIS AND DESIGN OF STRUCTURAL CONNECTIONS: Reinforced Concrete and Steel**
Hudson, J.A.	**ROCK ENGINEERING SYSTEMS: Theory & Practice**
Iyengar, N.G.R.	**STRUCTURAL STABILITY OF COLUMNS AND PLATES**
James, A.N.	**SOLUBLE MATERIALS IN CIVIL ENGINEERING**
Jardine, R.J. & Hight D.W.	**SOFT CLAY FOUNDATION ENGINEERING**
Jewell, R.A., Houlsby G.T., Milligan G.W.E.	**REINFORCED SOIL**
Jordaan, I.J.	**PROBABILITY FOR ENGINEERING DECISIONS: A Bayesian Approach**
Kutmanova, I., Skaloud, M., Tochacek, M.	**PLASTIC DESIGN IN WEBS, PLATES AND SHELLS**
Kwiecinski, M.	**PLASTIC DESIGN OF REINFORCED SLAB-BEAM STRUCTURES**
Lencastre, A.	**HANDBOOK OF HYDRAULIC ENGINEERING**
Leroueil, S., Magnan, J.P. & Tavenas, F.	**EMBANKMENTS ON SOFT CLAYS**
MacLeod, I.	**ANALYTICAL MODELLING OF STRUCTURAL SYSTEMS**
Megaw, T.M. & Bartlett, J.	**TUNNELS: Planning, Design, Construction**
Megaw, T.M. & Bartlett, J.	**TUNNELS: Planning, Design, Construction**
Megaw, T.M. & Bartlett, J.	**TUNNELS: Planning, Design, Construction, Volume 2**
Melchers, R.E.	**STRUCTURAL RELIABILITY ANALYSIS AND PREDICTION**
Mrazik, A., Skaloud, M. & Tochacek, M.	**PLASTIC DESIGN OF STEEL STRUCTURES**
Palossy, L., Scharle, P., Szalatkay I.	**EARTH WALLS**
Pavlovic, M.	**APPLIED STRUCTURAL CONTINUUM MECHANICS: Elasticity**
Skaloud, M.	**STEEL STRUCTURE STABILITY: Design & Applications of Webs & Flanges**
Spillers, W.R.	**INTRODUCTION TO STRUCTURES**
Vilnay, O.	**CABLE NETS AND TENSEGRIC SHELLS: Analysis and Design Applications**
Williams, K.	**CIVIL ENGINEERING CONTRACTS VOLUME 1: The Law, Administration & Safety**
Williams, K.	**CIVIL ENGINEERING CONTRACTS VOLUME 2: Planning, Estimating & Tendering**
Zloković, G.	**GROUP THEORY AND G-VECTOR SPACES IN STRUCTURAL ANALYSIS: Vibration, Stability and Statics**

ROCK ENGINEERING SYSTEMS
Theory and Practice

JOHN A. HUDSON, Ph.D., D.Sc., C.Eng.
Rock Engineering Consultants and
Imperial College, University of London

ELLIS HORWOOD
NEW YORK LONDON TORONTO SYDNEY TOKYO SINGAPORE

First published in 1992 by
ELLIS HORWOOD LIMITED
Market Cross House, Cooper Street,
Chichester, West Sussex, PO19 1EB, England

A division of
Simon & Schuster International Group
A Paramount Communications Company

Printed and bound in Great Britain
by Bookcraft, Midsomer Norton

British Library Cataloguing in Publication Data

A catalogue record for this book is available from the British Library

ISBN 0–13–782624–9

Library of Congress Cataloging-in-Publication Data

Available from the publisher

Table of Contents

Preface

The purpose of this book is to describe a new systems methodology with potential application to all rock engineering problems. From the outset, the approach treats rock engineering as a complete system. In other words, the approach starts with a 'top-down' analysis of the total system, rather than a 'bottom-up' synthesis of rock mechanics and rock engineering models. The methodology potentially encompasses the entire range of rock engineering systems and sub-systems, from the thermodynamics underpinning right through to the choice of appropriate contractual procedures.

The theory was developed to fulfil a need. As more technical information becomes available, as rock engineering projects become more complicated, as environmental issues become more significant and as more emphasis is placed on quality assurance, we clearly need some form of overview methodology that can provide procedural guidance in all circumstances. In the course of the work leading up to the book, many projects have been visited and many practitioners have told me what they require. The theory and associated methodology described here are the results.

This type of research development and the related implementation of the methodology into engineering practice are not possible without financial and moral support over many years. I particularly wish to thank the US and UK Governments — they have not only provided me with financial support, but have also encouraged me to pursue the ideas that are described here. Thus, the book is primarily dedicated to these two Governments.

To develop new ideas, some form of isolation is also needed. I found this isolation in the countryside of the Wye valley on the Welsh–English border, where many happy days have been spent watching the theory grow on the pages of my notebooks. In fact, one of the sayings of the farmers living in that area summarizes the thrust of this book: in Welsh, "Arferiad aeth heibio, anturiaeth sy' ddod";
and in English, "Behind is custom; ahead is adventure".
Therefore, the secondary dedication is to the countryside in the Wye valley.

In addition to having this long term support and beautiful countryside in which to think, I have been lucky to have been surrounded by intelligent and committed colleagues who have stimulated my thoughts and helped to reduce the number of errors in the book. Thanks are extended to everyone who has discussed interaction

Dedication

This book is dedicated to:

The US and UK governments for their support over the last 26 years

and to

the beautiful countryside on the Welsh-English border near Hay-on-Wye where many of the ideas in this book were generated and developed.

matrices and rock engineering systems with me — in the Wye valley, at Imperial College and elsewhere — especially Peter Arnold, Christine Cooling, Lyn Flook, Kemal Gokay, John Harrison, Carol Hudson, Fin Jardine, Dean Millar, Max de Puy Doug Spencer, Akio Tamai, Branko Vukadinovic and Jiao Yong.

Production of this book has been facilitated by the advent of computers. The diagrams were computer drawn by Miles Hudson, Willison Mutagwaba, Branko Vukadinovic and Wu Bailin. The raw text was typed by Linda Moore from dictation, and the final camera-ready manuscript prepared by the author and his wife, Carol, using the Aldus PageMaker program on an Apple Macintosh computer. I should like to thank my publishers, Ellis Horwood, and Acquisitions Editor, Karen Rose, for their rapid reaction to my request that this book should be written and published to a strict schedule.

The book is intended for everyone involved in rock engineering: clients, consultants, contractors, students, teachers and researchers. Most of the methodology has been successfully field tested; however, some additional and perhaps more speculative ideas have been included in Chapters 6 and 8 to indicate directions that future developments might take.

I have tried to convey the concepts with one-hundred per cent reader comprehension; accordingly, many of the ideas are illustrated by direct reference to photographs of rock engineering activities. Even so, if there is anything presented in the following chapters that you do not understand, it is my fault.

J A Hudson, 1992

Rock Engineering Consultants
7 The Quadrangle
Welwyn Garden City
Herts AL8 6SG, UK

Imperial College of Science,
Technology & Medicine
University of London
London SW7 2BP, UK

Photograph acknowledgements

The author is indebted to all the various organizations and individuals who made the use of the photographs in this book possible. Acknowledgements, where appropriate, are included below — with thanks.

Front cover	Redland Aggregates Ltd, UK
Fig. 1.1	Woodmansterne Ltd, UK
Fig. 1.2	Woodmansterne Ltd, UK
Fig. 1.3	Redland Aggregates Ltd, UK
Fig. 1.4	Laboratorio Nacional de Engenharia Civil, Portugal
Fig. 1.7	Imperial College, University of London, UK
Fig. 2.3	Transport & Road Research Laboratory, UK
Fig. 2.5	Underground Research Laboratory, AECL Ltd, Canada
Fig. 2.9	University of Minnesota, USA
Fig. 2.18	Sachtleben Bergbau GmbH, Metallgesellschaft AG, Germany
Fig. 2.19	Redland Aggregates Ltd, UK
Fig. 2.21	Transport & Road Research Laboratory, UK
Fig. 4.1	Redland Aggregates Ltd, UK
Fig. 4.2	Imperial College, University of London, UK
Fig. 4.10	Transport & Road Research Laboratory, UK
Fig. 4.19	Sezai Turkes Feyzi Akkaya Construction Co, Turkey
Fig. 4.20	University of Wisconsin, USA
Fig. 5.4	Professor M Romana, Spain
Fig. 5.7	Panama Canal Commission
Fig. 5.8	Panama Canal Commission
Fig. 5.11	Angelsey Mining plc, UK
Fig. 5.12	British Coal, UK
Fig. 5.13	INCO Ltd, Canada
Fig. 5.15	University of Wisconsin, USA
Fig. 6.5	British Museum (Natural History), UK
Fig. 6.6	British Museum (Natural History), UK
Fig. 6.9	Transport & Road Research Laboratory, UK
Fig. 6.12	Transport & Road Research Laboratory, UK

Fig. 7.4	Sir William Halcrow & Partners Ltd, UK
Fig. 7.8	Camborne School of Mines, UK
Fig. 7.10	Redland Aggregates Ltd, UK
Fig. 7.12	Redland Aggregates Ltd, UK
Fig. 7.14	Transport & Road Research Laboratory, UK
Fig. 7.18	Building Research Station, UK
Fig. 7.19	University of Trabzon, Turkey
Fig. 7.21	Redland Aggregates Ltd, UK
Fig. 8.3	Transport & Road Research Laboratory, UK
Fig. 8.9	Building Research Station, UK
Fig. 8.15	Japanese Government
Fig. 8.16	Japanese Government
Fig. 9.2	Redland Aggregates Ltd, UK
Fig. 9.3	Laboratorio Nacional de Engenharia Civil, Portugal
Fig. 9.8	Transport & Road Research Laboratory, UK
Fig. 9.9	Laboratorio Nacional de Engenharia Civil, Portugal
Fig. 10.8	National Central University, Taiwan
Fig. 10.9	National Central University, Taiwan
Fig. 10.13	Underground Research Laboratory, AECL Ltd, Canada
Fig. 10.17	NASA, USA
Fig. 10.18	National Geographic Society, USA
Back cover (matrix)	Kemal Gokay, Imperial College, University of London, UK

1

Introduction

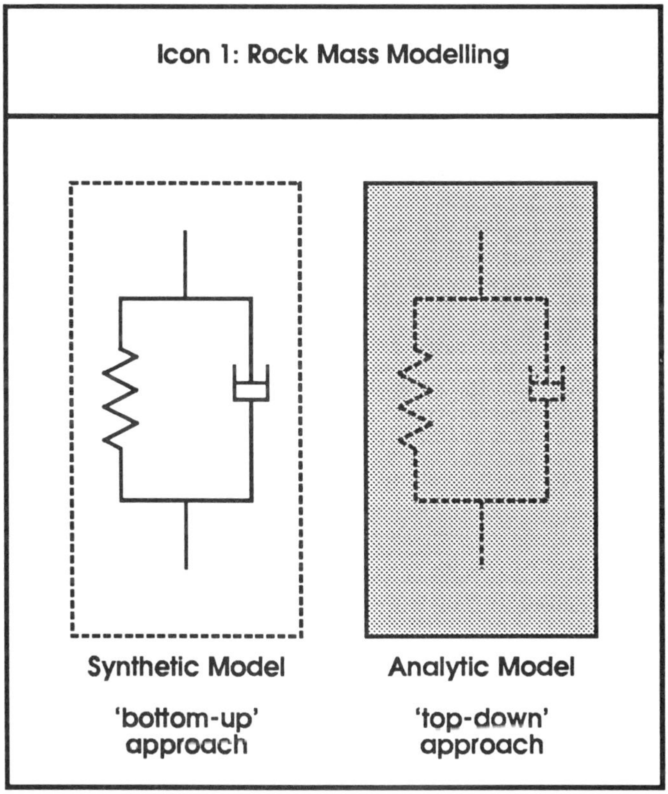

In Chapter 1 the reasons for adopting a systems approach are discussed. Icon 1 consists of the two main approaches to rock mass modelling: the 'bottom-up' synthetic model on the left; and the 'top-down' analytic model on the right.

1.1 ROCK ENGINEERING BACKGROUND

We know that mankind has been involved in rock engineering for thousands of years, and it is likely that this branch of engineering extends back hundreds of thousands of years: its history is lost in the mists of time. We have evidence of very early structures, both carved out of rock and constructed with rock blocks. Some of these structures are crude; others are a testament to the skills of the prehistoric engineers.

In Fig. 1.1, there is a photograph of Stonehenge, a structure of unknown purpose (probably astronomical) built about five thousand years ago in the south of England. Despite the simplicity of this structure, considerable skills were required to excavate the rock and construct the trilithons visible in the picture. The sophistication of

Fig. 1.1 The rocks of Stonehenge, a prehistoric structure built on Salisbury Plain in the UK five thousand years ago.

structures built from rock is evident in Egyptian, Greek and Roman buildings and reached a magnificent zenith during the major cathedral building era in Europe.

The interior of Wells Cathedral in England is shown in Fig. 1.2. This is the first building in England where the Gothic pointed arch, in association with the compound Gothic pillar, was used exclusively. The Cathedral, finished in 1230, dramatically illustrates the skills of the architects and stonemasons of almost seven hundred years ago. The scissor arches, clearly visible in the photograph, are of structural and geotechnical interest because they are not part of the original structure. About two hundred years after completion of the cathedral, the central tower began to collapse because of foundation and structural problems and the scissor arches were added to stabilize the structure. The sweeping lines of these arches could well have emerged from a stress analysis carried out today.

Although there is this long history of rock engineering over thousands of years,

Fig. 1.2 The scissor arches in Wells Cathedral, UK.

rock mechanics as a scientific subject in its own right has only been developed over the last thirty years or so. All of the earlier activity was, of course, conducted without the benefit of modern knowledge. In some cases the projects were successful, often dramatically so, as illustrated in Figs. 1.1 & 1.2; but, in other cases, we know that they were unsuccessful. Many cathedrals were not so fortunate as that at Wells and collapsed during or shortly after construction.

Now we have the benefit of at least thirty years development in rock mechanics technology *per se*. We also have the benefits of its implementation in a whole variety of rock engineering activities, encompassing civil, mining and petroleum engineering. Often, this knowledge is utilized fully in the design, construction and monitoring of engineered structures. Sometimes, the knowledge is not used for a variety of reasons.

Firstly, rock masses are natural: rock masses are not fabricated according to the project requirements; the properties have to be established by site investigation. Moreover, rock masses are fractured, often in complex ways. This is illustrated by the picture in Fig. 1.3 which shows a 20 m high bench in a quarry in the Mountsorrel granodiorite in the UK. The appearance of the rock immediately indicates the need for special techniques to account for the fractures in any engineering design for construction on or in this rock. The complexity of this degree of fracturing can be quite daunting. The fact that the rock cannot be completely described has caused engineers to abandon design techniques 'borrowed' from structural analysis used with man-made materials.

Secondly, the relevant technical information does exist in the main twenty-five or so published textbooks and in experts' minds, but it is often not easy for those engaged in construction — clients, consulting engineers and contractors — to tap this expertise in any coherent fashion.

The third and last factor to consider, especially in this final decade of the twentieth century, is that rock engineering is becoming more complex. More mining is being considered at deeper levels, larger caverns are being envisaged and there are many new applications for which there is no precedent practice. For these latter projects,

Fig. 1.3 Granodiorite quarry bench — illustrating the fractured nature of rocks, Mountsorrel Quarry, UK.

we can look on the drawing board (or computer screen) to study the proposed projects We then note that there are, for example, no geothermal energy sources currently in production with outputs matching those of conventional electricity generating systems. There is no city located totally underground. In the context of radioactive waste disposal, no one has yet constructed, emplaced waste and sealed a repository. Yet these are the types of construction problems with which we will be faced.

An approach to rock engineering is now required that explicitly takes into account the very special nature of rock as an engineering material, inherently incorporates the current knowledge base, and can accommodate any rock engineering project objective. The approach must also allow for the pressures being imposed by financial and environmental circumstances. We naturally wish to conduct rock engineering

Fig. 1.4 Illustration of the powerhouse cavern in the Alto Lindoso hydroelectric scheme in Portugal.

in an optimal way but it must be borne in mind that this not only includes achieving the engineering objective at the minimum cost but also minimizing disturbances to the natural environment. Furthermore, there is increasing awareness of the need to ensure that what is supposed to have been done has actually been done, i.e. to have good quality assurance procedures and auditing mechanisms.

In Fig. 1.4 there is a picture showing construction of a hydroelectric powerhouse cavern in Portugal. Not only must this cavern be constructed adequately in its own right, but its function must also interface successfully with the rest of the scheme. In other words, the complete system of rock engineering must be taken into account — with all its main and subsidiary objectives.

1.2 PURPOSE OF THE BOOK

The aim of this book is to present an approach to rock engineering which addresses all the points in the previous Section. The methodology is new and has been developed by the author in direct response to the need, widely expressed by practitioners, for an 'all-encompassing' procedural technique to approach increasingly complex rock engineering problems. The approach is an objective-based methodology and is capable of utilizing all the existing information relevant to a particular project and of tailoring the procedures to the circumstances. In short, the aim of the book is to present a new Rock Engineering Systems (RES) approach to rock engineering.

Traditionally, the development of models representing rock behaviour has evolved from the simple to the complex. For example, we might assume that the mechanical behaviour of a rock mass can be approximated by a spring. We then find that there is some time-dependency and so we add a dashpot to the model. This is the synthetic model shown on the left hand side of Icon 1 at the beginning of this Chapter and on the left hand side of Fig. 1.5. We find further discrepancies between the model and observations, and continue to amend the model via a synthetic modelling procedure.

Thus, with reference to Fig. 1.5, traditional rock mass modelling has been conducted via the 'exact representation' approach. The components shown on the left side of Fig. 1.5 have exact mathematical characteristics. The model is synthetic because it is built up from the components. It represents a part of the rock mass and has an unknown boundary of applicability.

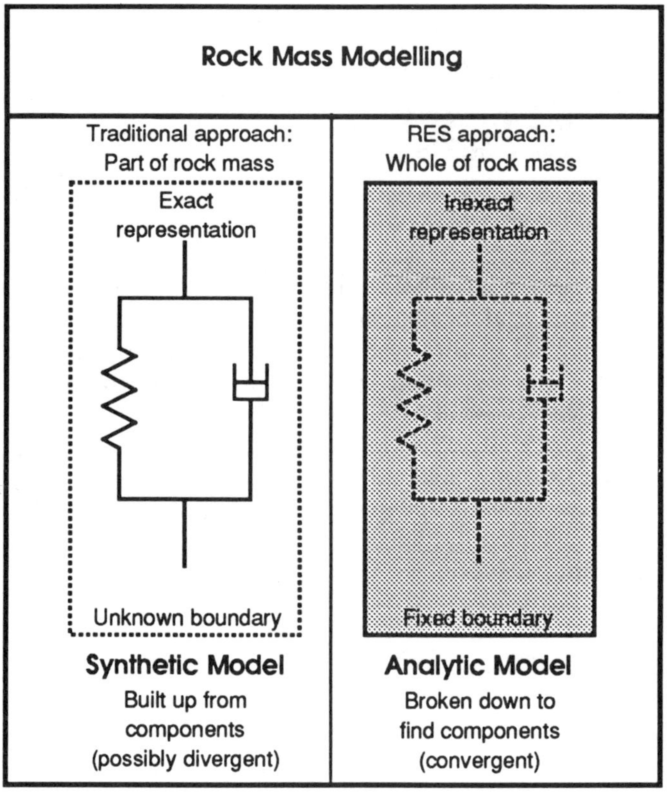

Fig. 1.5 Expanded version of Icon 1 (on first page of Chapter 1), emphasizing the difference between the two main modelling approaches (the traditional and RES approaches).

The thesis of this book is that the rock engineering projects and associated science are now so complex that this 'bottom-up' approach becomes too complicated and, more importantly, will not (in the general case) be convergent to the correct model. There is no aspect of this synthetic modelling procedure that will ensure all the links are present: indeed, such a synthetic model can be completely invalidated by omission of a critical component. Also, the synthetic model does not comfortably take into account the wider perspective of rock engineering — i.e. all the objectives, interfaces and constraints mentioned earlier.

The approach adopted here is a total systems approach represented by the analytic model in the right-hand sides of Icon 1 and Fig. 1.5. The model is termed 'analytic' because it is assumed that the whole of the rock mass and its characteristics are contained within the solid border line — and then we analyze the system. (Note the dotted border line of the synthetic model representing the unknown boundary, both conceptually and physically.) So, the idea of the analytic model is that we define everything to be present in the model: the problem is simply to find out what is there. In this way, all the necessary information can be utilized and the rock mechanics component put into its correct context in rock engineering. Naturally, the representation of the mechanical mechanisms will be inexact in the analytic model when compared to the synthetic model. Thus, in both Icon 1 and Fig. 1.5, the alternation of the dashed and solid lines representing the boundary, and the specific internal model directly reflects the characteristics of the modelling approach.

The two terms 'synthetic' and 'analytic' have been used by various authors to mean different things. The terms are used througout this book according to their dictionary meanings: 'synthetic' for building up; and 'analytic' for breaking down.

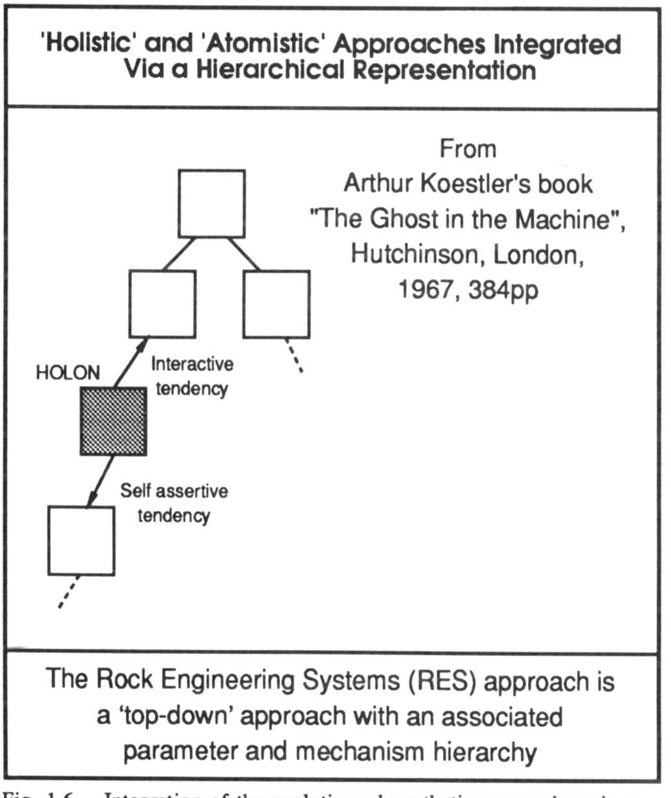

Fig. 1.6 Integration of the analytic and synthetic approaches via a hierarchical representation.

Although the history of rock mechanics has to date followed the synthetic model approach, there is, of course, nothing new in other subjects about the concept of the analytic model which begins with an overview of the system. Two other words used to describe the synthetic and analytic models are 'atomistic' and 'holistic' respectively. The distinction between these two approaches in science and engineering has been the subject of philosophical debate for many centuries and it is not necessary to summarize here all the points of view that have been expressed. The reason for adopting the top-down, or analytic, or holistic approach is simply that it has become too difficult to utilize the synthetic approach, given the points discussed in Section 1.1.

A method is required whereby we have a list of all rock properties and an understanding of all rock mechanisms as our fundamental knowledge base. We also need to know precisely what it is we are trying to do: in other words, the method should be objective-based. We then need a procedure for identifying the mechanisms and rock properties relevant to our project, within the context of the objective — and hence the ability to establish the relevant engineering techniques. In this way, we will utilize existing knowledge in an optimal way to develop site investigation, design, construction, and monitoring procedures for any project. The presentation of this Rock Engineering Systems (RES) approach is the purpose of the book.

It should be noted that to use only either the synthetic or analytic models is an unnecessary constraint: they both have their advantages and disadvantages. We will start with the analytic model and see how far it takes us, but incorporate the advantages of the synthetic approach whenever helpful. In Fig. 1.6 there is a diagram showing one of the ideas of Arthur Koestler (see reference in Fig. 1.6) to integrate the atomistic and holistic approaches via a hierarchical representation. He termed a particular position in the hierarchy a 'holon': if one considers the part this

Fig. 1.7 Major slide occurring at the Teutonic Bore open pit mine in Western Australia.

holon plays in the scheme of things by looking upwards in the hierarchy, an interactive tendency is being invoked; conversely, if one considers how the holon controls the remaining hierarchy beneath, a self-assertive tendency is being invoked. Although this idea may be more relevant to the life sciences, there are advantages in integrating the two approaches via this type of hierarchy. We will see later in the book how a parameter and mechanism hierarchy can be developed. We will also see how an individual parameter can be considered in two main contexts: the effect of the parameter on the system; and the effect of the system on the parameter.

Inherent in the systems approach is the ability to consider the conditions that can lead to catastrophic failure, i.e. disaster scenarios. Is the engineering system stable or unstable? How much would one parameter have to change to trigger catastrophic failure? What are the unacceptable combinations of parameters? All these questions can be considered once a systems approach has been established.

In Fig. 1.7, a major slide is illustrated at the open-pit Teutonic Bore mine in Western Australia. This collapse was predicted at the time from displacement measurements, but we would wish to be aware of such a possibility early on in the design process. The possibility of this sort of failure and indeed any conceivable type of failure can be identified and assessed by the systems approach.

1.3 STRUCTURE OF THE BOOK

The structure of the book follows the philosophy already outlined, each chapter being prefaced by an iconic representation of the key concept. In the next Chapter there is an explanation of the interaction matrix device developed by the author as the basic tool for the Rock Engineering Systems (RES) approach to rock engineering. The interaction matrix is a presentational technique and an analytic tool for simultaneously representing the key parameters (as leading diagonal terms) and the rock engineering mechanisms (as interactions or off-diagonal terms).

The extension of the interaction matrix to include concepts, sketches and photographs leads automatically to an Atlas of Rock Engineering Mechanisms — which is presented for slopes and underground excavations in Chapter 3. In order to study parameter interaction intensity and dominance in the Rock Engineering System (RES), it is necessary to code the matrix components. Once this has been achieved, the parameters can be plotted in system cause *vs.* effect space. This procedure, together with two associated theorems, is explained in Chapter 4. Matrix coding examples for slopes and underground excavations are given in Chapter 5.

The interaction matrix device shows binary mechanisms, i.e. a mechanism linking two parameters. In fact, we definitely expect ternary and higher level mechanisms to be operating, both concurrently and consecutively. It transpires that an extremely useful way of considering all the possibilities is via interaction matrix pathway analysis. The associated multiple rock engineering mechanism analysis is described in Chapter 6.

Having thus established the basis of the theory in Chapters 2–6, the systems understanding of rock engineering is reviewed in Chapter 7, including the interpretation of existing procedures and the link with conventional systems analysis.

In rock mechanics, there are often difficulties with establishing parametric properties at a point — because of the discontinuous nature of the rock and other factors. Thus, it is useful to consider the integrated nature of the properties via energy flux within the rock engineering system. This leads to energetic coding of the interaction matrix and consideration of energy sources, sinks and pathway flows. Also, the process of engineering can be interpreted entropically. For any order created by engineering, there will be a greater degree of disorder induced in the surrounding region. These factors are discussed in Chapter 8.

In Chapters 9 and 10, the implementation of the Rock Engineering Systems (RES) approach and 'systems thinking' are explained. In particular, the complete implementation is conducted, firstly, 'top-down' and, secondly, 'bottom-up' — combining the analytic and synthetic methods. The identification of key parameters and mechanisms is achieved from the knowledge infrastructure through a process known as 'hierarchical winnowing and matrix compaction', which allows the system and procedures to be tailored to any specific project objective or objectives.

The adoption of a systems approach is formalized through the R^2 scheme. R^2 is the acronym of the acronym REMIT/RESPONSE. This stands for "Rock Engineering Mechanisms Information Technology/Rock Engineering System Performance & Objective-based Network Sequence Evaluation". Using R^2, one enters the system analytically, and exits developing the appropriate procedures synthetically.

The whole Rock Engineering Systems (RES) approach is ideally suited for computerization, especially utilizing the newly developed computing techniques such as object-oriented programming and iconic interface multimedia programming. In the hope that their use will be not too far off, the link with neural networks is also explained.

In the last Chapter on systems thinking, questions are discussed such as: "What choices are available to the engineer to alter the system?"; "Is the engineer in control?"; and "What are the unacceptable combinations of parameters?". The answers to these questions lead automatically to the ability to evaluate candidate schemes via this systems approach and to be able to conduct strategic assessments of project robustness when evaluating candidate schemes.

Rock engineering audits are also discussed: these can be financial, environmental, energetic and entropic, of which the highest level is the entropic audit — assessing the degree of disorder induced by the engineering.

2

Interaction Matrices

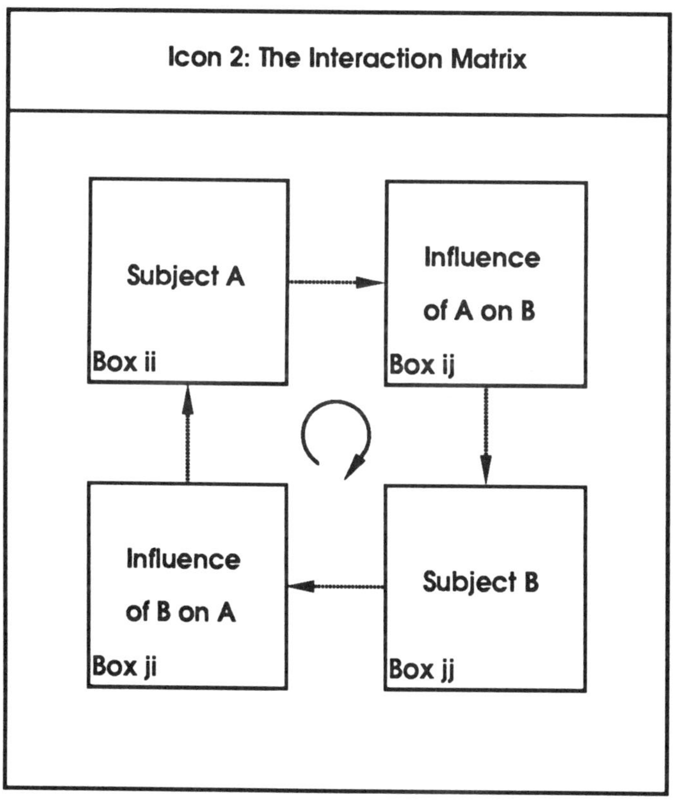

For the systems approach presented in this book, the device used to identify and link all the rock engineering mechanisms (and associated parameters) is the interaction matrix, shown in iconic form above.

2.1 THE NATURE OF INTERACTION MATRICES

Given that an analytic approach is being developed, we must have some method of representing the total system behaviour. Also, because it would be helpful to be able to utilize the same presentational and analytic procedures for any circumstances or project, there is obvious benefit in adopting a methodology that has universal applicability. The interaction matrix has been developed by the author for this purpose: the basic method of representing the relevant parameters, their interactions, and the rock mass/construction behaviour is via the interaction matrix — which is shown conceptually in Icon 2 on the previous page. The fact that potentially there are interactions between all things has long been recognized, as is illustrated by the lines of poetry in Fig. 2.1.

Interactions

All things by immortal power,
Near or far,
Hiddenly
To each other linkèd are,
That thou canst not stir a flower
Without troubling of a star.

Francis Thompson
English Victorian poet
(1859–1907)

Fig. 2.1 Part of a poem suggesting that there could be a relation between all things.

We are restricting our interests to rock engineering, which involves both natural interactions in the rock mass and the interactions induced by construction. As an example of the interactions in a discontinuous rock mass, six are shown in Fig. 2.2. In this Figure there are three basic subjects being considered: the rock structure, the rock stress and water flow. The number of binary combinations of these three parameters is six, e.g. the six interactions shown in the Figure.

In Interaction 1, the rock structure (i.e. the discontinuities) affects the local stress field because the principal stresses are rotated and altered in magnitude. In the extreme case, if the discontinuity is open and hence can transfer no stress, the principal stresses must be parallel and perpendicular to the discontinuity surfaces. Thus, although the far-field boundary conditions may be considered in terms of an overall average stress value, the local stress field in the proximity of discontinuities may be completely different.

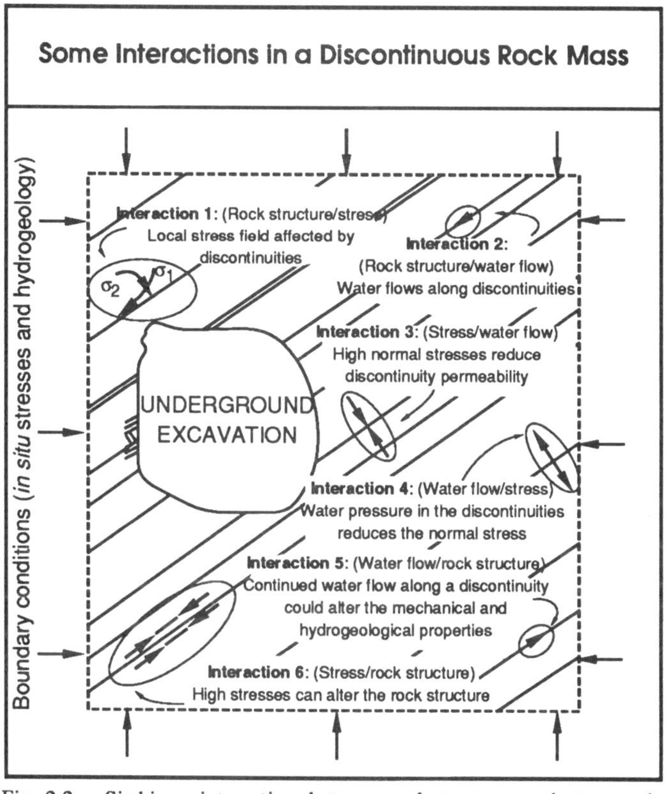

Fig. 2.2 Six binary interactions between rock structure, rock stress and water flow in the region around an underground excavation.

In Interaction 2, we consider how the rock structure affects the water flow. The primary permeability is a function of the intact rock; the secondary permeability is a function of the fractures. In the context of the typical duration of engineering construction, the secondary permeability is usually far more significant than the primary permeability. Thus, it is the discontinuities that dictate the permeability and hence the water flow that occurs in the rock mass and into the excavation.

In Interaction 3, we consider how the stress might affect the water flow. Because the water is flowing through the discontinutities, any closure induced by stress is likely to have a significant effect on the water flow. Indeed, this is the case because the water flow is often assumed to be proportional to the cube of the aperture and very little stress is required to alter the aperture when the discontinuity is fairly open.

In Interaction 4, the complementary effect of the influence of the water flow on the stress is illustrated. Water pressure in the discontinuities reduces the normal stress. This effect is well known, and there is the concept of effective stresses to assist in the analysis of this interaction.

In Interaction 5, we consider how the water flow might affect the rock structure. We have already seen in Interaction 2 the dramatic effect that the rock structure has on governing the water flow. But how does the water flow affect the rock? What is the complementary effect? Continued water flow along a discontinuity is highly likely to alter the mechanical and hydrogeological properties of the rock mass. We noted in the previous paragraph that for Interaction 4 there is a whole body of knowledge, i.e. the use of effective stresses, to account for the interaction. Conversely, however, for Interaction 5 much less is known about the subject.

Finally, in Interaction 6, we consider the complementary effect to Interaction 1. How does the stress affect the rock structure? The very presence of the discontinuities has resulted from orogenic and other geological disturbances over long periods of time, i.e. through stress and strain changes in the rock mass.

With just these six interactions, shown in Fig. 2.2 and discussed above, it is clear that some methodological device is needed to be able to coherently consider the interactions — and indeed any other interactions that may be occurring. Also, not only must the natural processes that are involved in the rock mass interactions be considered but also the effects of construction. This refers both to the disturbance induced by the construction process and the presence of the new structure created by the construction. Naturally, we might expect the construction of an underground

Fig. 2.3 'Saw tooth' tunnel wall caused by the interaction between master joints in the rock mass and excavation by blasting, Birkbeck tunnel beneath the Cumbrian hills, UK.

excavation to alter the stresses and hence to alter the apertures of proximate discontinuities, and perhaps even to cause slip on existing discontinuities and to create new discontinuities. An example of the effect of the rock mass structure on the construction process is shown in Fig. 2.3, which is a photograph of the tunnel constructed to take water from the Lake District to Manchester below the Cumbrian hills in the UK. The rock mass shown contains a set of master joints which have influenced the tunnel blasting. The 'saw tooth' form of the tunnel wall is a direct manifestation of the interaction between the rock mass structure and the construction method.

As mentioned, the method of presenting and considering the interactions is to use an interaction matrix — see Icon 2 at the introduction to this Chapter. The subjects or parameters in question are placed along the leading diagonal (top left to bottom right) of a matrix. In Icon 2, this is a 2x2 matrix: Subject A is in the top left-hand box and Subject B is in the bottom right-hand box. The influence of A on B is located in the top right-hand box; the influence of B on A is located in the bottom left hand box. Thus, the basic principle of the interaction matrix is to list the main subjects or parameters along the leading diagonal and to consider the interactions in the off-diagonal boxes.

In a 2x2 matrix there are two leading diagonal terms and two off-diagonal terms, as is shown in Icon 2 on the first page of this Chapter. In a 3x3 matrix there will be three leading diagonal terms and six off-diagonal terms — for example the six interactions that we discussed in reference to Fig. 2.2 and which would have leading diagonal terms <u>rock structure</u>, <u>rock stress</u> and <u>water flow</u>. In a 12x12 matrix, there would be twelve leading diagonal terms and 132 off-diagonal terms. Under the latter circumstances, and indeed with any significant number of leading diagonal terms, the presentational technique of the interaction matrix is not only convenient but also becomes essential. Unless one has a device of this kind for identifying the multiplicity of interactions, it is highly unlikely that they will all be taken into account.

At this stage in using the interaction matrix, only binary interactions are being considered; later, methods of studying compound mechanisms involving any number of parameters will be introduced. Similarly, in such an analysis with ternary and higher level interactions, it is absolutely essential to have a coherent method of identifying the interactions and the part that they play in the total system.

To illustrate how two of the interactions shown in Fig. 2.2 can be accommodated by this technique, a 2x2 matrix is shown in Fig. 2.4. The two leading diagonal terms are the <u>rock discontinuities</u> and the <u>rock stress</u>. In the top right-hand box, we can see the influence of a discontinuity on the rock stress: an open discontinuity affects the stress field, and in particular alters the principal stresses. In the bottom left-hand box, we see how the rock stress affects the rock discontinuities in terms of the aperture closure.

There are several important points to be noted in this simple 2x2 matrix in Fig. 2.4. Firstly, the leading diagonal terms of the matrix can be subjects or concepts: they do not have to be quantitative values — although later on in the book we will see that there are advantages in ensuring that the 'quality' of the leading diagonal terms is the same, e.g. they could all be expressed in energy units. Secondly, the

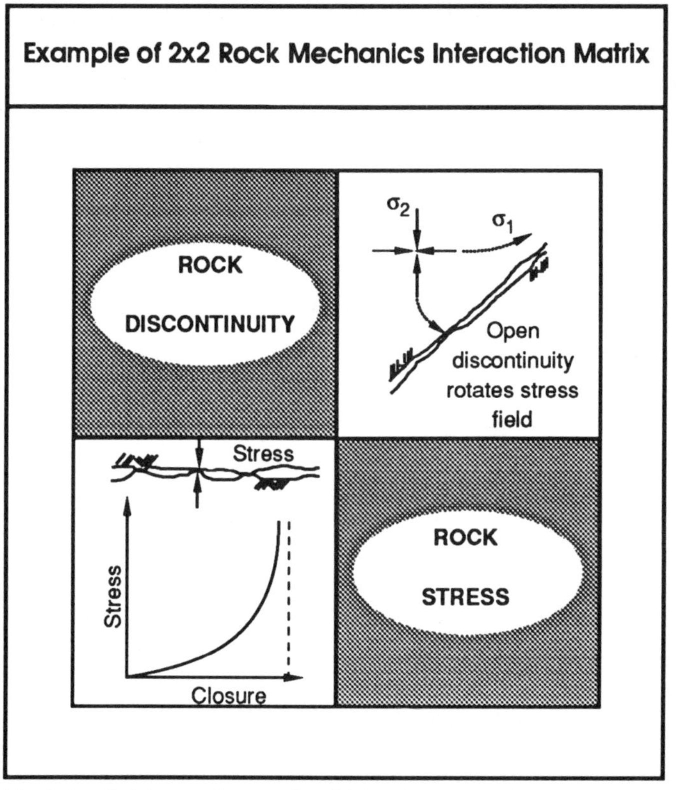

Fig. 2.4 2x2 interaction matrix with leading diagonal terms, Rock Discontinuity and Rock Stress.

two off-diagonal terms are not the same: the way in which the rock discontinuity influences the rock stress is not the same as the way in which the rock stress influences the rock discontinuity. Thirdly, it is clear that we could increase the basic 'resolution' of the matrix by considering various characteristics of the rock discontinuity and the rock stress. For example, we could divide the rock discontinuity into discontinuity orientation and discontinuity aperture. Similarly, we could divide the rock stress into principal stress directions and principal stress magnitudes. Using these four characteristics as the leading diagonal terms of the matrix, we could create a 4x4 matrix which would have 12 off-diagonal terms — obviously resulting in a deeper understanding of the mechanisms involved.

There are various possibilities for presenting information in the off-diagonal boxes. From the discussion in the previous paragraph, an interaction box could contain a major analysis of the mechanisms involved. Alternatively, we might try to represent the significance of the mechanism by some form of coding device. We could even show a photograph of the phenomenon. All of these have their advantages in different circumstances. Note, however, that the interactions illustrated in Figs. 2.2 and 2.4 are examples of the influences and are not comprehensive: they serve as illustrative examples in explaining how the matrices are developed.

Fig. 2.5 Core 'discing' that occurs when rock core is obtained from highly stressed rock — these are the 'Manitoba biscuits' obtained at the Underground Research Laboratory, AECL, Canada.

An example of a photographic presentation is shown by the photograph of core discing that occurs when diamond drillholes are made at the Underground Research Laboratory in Canada (Fig. 2.5). This is a perfect example of the influence of construction on the intact rock or, with reference to Fig. 2.4, the influence of rock stress (in this case induced) on rock discontinuities (in this case actually created by the rock stress).

Fig. 2.6 Illustration of a rock mass with very variable discontinuity characteristics — indicating the need for careful consideration of the 'resolution' of the mechanisms.

It is apparent from the foregoing discussion that there are many interactions to be considered and that we must initially ensure that the interaction matrix edifice that is created will be able to deal comprehensively with any circumstances. In Fig. 2.6 there is a photograph of a rock mass containing different types of discontinuities: some are persistent and open; some are impersistent and closed. How do we account for the multi-faceted nature of the rock mass, all the mechanisms that might be operating now, and all the mechanisms that might operate both during and after construction?

In Fig. 2.7, the leading diagonal of the interaction matrix contains the three terms provided by the rock mass itself, rock structure, *in situ* stress, water flow, and the fourth term, construction. Although this matrix only has dimensions 4×4, there are still 12 off-diagonal terms — and it would be difficult to be sure of considering

Rock Mechanics - Rock Engineering
4x4 Interaction matrix

Rock Structure F_{ij}	Fractures affect the values and orientations of the stresses	The fracture network governs the secondary permeability	Fractures can influence the size and orientation of excavations
Stesses can open or close fractures, and also create them	**Rock Stress** σ_{ij}	In general, the higher the normal stress, the lower the permeability	High rock stresses can cause construction failures
Continual water flow in fractures affects their properties	Normal stresses reduced by water pressure	**Water Flow** K_{ij}	Grouting and drainage may be required during construction
Blasting can damage old fractures and create new fractures	In the vicinity of excavations the principal stresses are altered	An excavation will always become a sink for the water flow	**Const-ruction** C_{ij}

The top left 3x3 matrix represents rock mechanics.
When the extra leading diagonal term Cij is added,
the resulting 4x4 matrix represents rock engineering.

Fig. 2.7 Interaction matrix with four leading diagonal terms: Rock Structure, Rock Stress, Water Flow, and Construction.

all these for a particular design, unless they were identifiable through the matrix or a similar device. Note that with the introduction of Construction as the last leading diagonal term, the architecture of the matrix can be interpreted in engineering terms. The right-hand column components represent the influence of the three rock mechanics parameters on construction and thus they will be important in design.

The lower row represents three ways in which construction affects the rock mass parameters, i.e. they are results of construction and can be monitored and used for back analysis. Note that all complementary off-diagonal terms in the matrix in Fig. 2.7 are different — and so the matrix is fully asymmetric.

In view of our observations on the nature of the interaction matrices, let us now consider the question of matrix symmetry and establish exactly why a matrix is symmetric or asymmetric and later go on to study matrix resolution. Both these aspects are crucial to the development of the theory and the understanding of rock engineering systems.

2.2 MATRIX SYMMETRY AND ASYMMETRY

In Fig. 2.8, the concepts of symmetry and asymmetry are illustrated by 2x2 stress–strain matrices. The two leading diagonal terms are stress and strain. Using the convention of a clockwise rotation, and with reference to the top matrix of Fig 2.8, the upper right-hand box of this matrix illustrates the influence of stress as an independent variable in causing strain. The lower left-hand box represents the

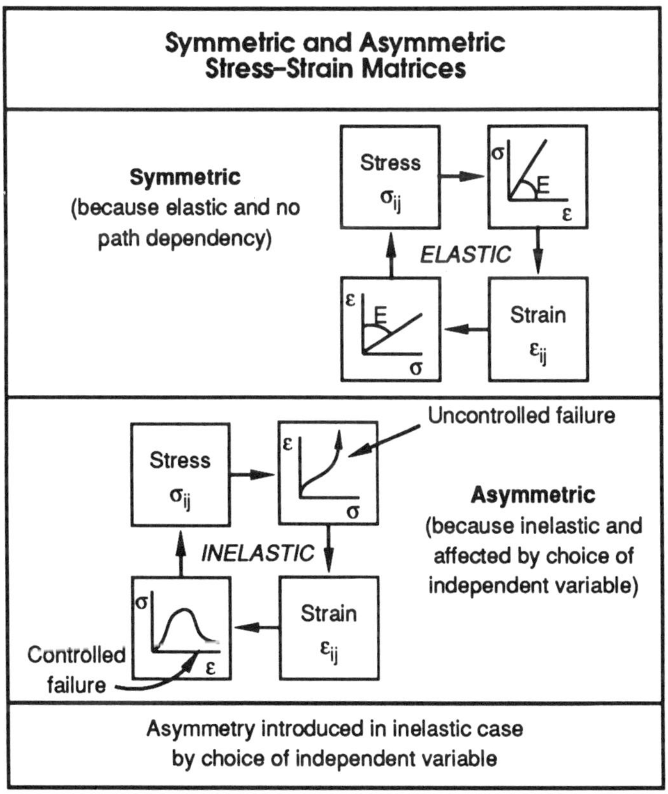

Fig. 2.8 The relation between stress and strain for the elastic and inelastic cases — illustrating matrix symmetry and asymmetry.

influence of strain as an independent variable in causing stress. Note that the two sketches in the off-diagonal boxes in the top matrix of Fig. 2.8 are identical except that, according to scientific convention, the independent variable is plotted on the x-axis in each case. Young's modulus, E, is shown in the diagrams although, of course, in the general case we would have to consider the generalized Hooke's Law with twenty-one elastic constants.

These diagrams have been plotted differently to highlight the cause and effect in each case. A matrix is symmetric when the complementary off-diagonal boxes are the same, i.e. when Box ij = Box ji. Indeed, the information in each box is the same — and so this matrix is symmetric. The top matrix in Fig. 2.8 represents elasticity and so this matrix would be expected to be symmetric.

However, the top 2x2 matrix in Fig. 2.8 should be contrasted with the bottom 2x2 matrix in the same Figure. In the bottom matrix, the stress and strain are continued into the rock failure region. Depending on the independent variable, the rock exhibits different behaviour, as shown by the graphs in the off-diagonal boxes. If stress is the independent variable, as shown in the top right-hand box of the bottom matrix in Figure 2.8, the inevitable consequence of a steady increase in the stress is uncontrolled failure at the compressive strength. Conversely, if strain is the independent variable and a constant strain rate is applied to the rock, the failure is usually controlled and occurs as a continuous structural breakdown process, as indicated by the lower left-hand box of the bottom matrix.

The bottom matrix in Fig. 2.8 is not symmetric: the influence of stress in causing strain is not the same as the influence of strain in causing stress, i.e. in the failure region. The top matrix in Fig. 2.8 is symmetric because the rock has remained elastic and there are no path-dependent effects. The bottom matrix is asymmetric

Fig. 2.9 Marble test specimen, 25mm in diameter (i.e. across the section shown above), that has been uniaxially loaded in compression throughout the complete stress–strain curve to destruction.

because, when the rock becomes inelastic, its behaviour critically depends on the choice of independent, i.e. controlled, variable.

The consequences of symmetry and asymmetry in the interaction matrices are far reaching and will be discussed further in Chapter 4. It is probably clear already that the rock engineering interaction matrices we will be generating will not be symmetric. A review of the interactions illustrated in Figs. 2.2 and 2.7 will reinforce this point: none of the complementary interactions are the same.

Another factor which we alluded to in reference to Fig. 2.4 was the level of complexity at which we could consider the problem. This is also illustrated with reference to Fig. 2.8. Although the lower left off-diagonal term of the bottom matrix in the latter Figure is simply represented as a complete stress–strain curve, this single curve reflects a whole host of individual mechanisms operating within the rock microstructure. In Fig. 2.9, a longitudinal section of a rock specimen at the conclusion of a controlled failure test is shown. At what level do we consider the mechanisms here? Is the complete stress–strain curve, as included in the lower left box of the bottom matrix in Fig. 2.8, too fine or too coarse? The single curve might be too fine a representation because, in a discontinuous, inhomogeneous, anisotropic, and time-dependent rock, there will be wide variations in the curve and we might only be interested in, say, the energy represented by the area beneath the curve. On the other hand, a single curve could be considered too coarse because we have not identified the individual mechanisms that contribute to the composite mechanisms represented by this curve. The general subject of matrix resolution will be taken up in Section 2.2.

To complete this Section on the principle of the interaction matrices, three mathematical matrices are presented to highlight the method of constructing the matrix and the associated symmetry/asymmetry considerations. In Fig. 2.10, a four-sided figure shape conversion matrix is presented. This was constructed by Tamai (pers. comm., 1990). The leading diagonal terms of the matrix are a square, a rectangle, a rhombus and a parallelogram. In each off-diagonal box, there is a 2x2 condition icon relating to the sides and angles. Thus, this interaction matrix shows how each of the four-sided figures can be converted to any of the others by altering the conditions relating to the sides and the angles. For example, a square can be converted to a parallelogram using the top right hand 2x2 icon: if we release the constraint of the square having four equal sides and four equal angles, we have a parallelogram — and similarly for all the other conversions. The link with set theory is intimated by the Venn diagram at the top right of the figure, but the full development of the link between the interaction matrices and set or group theory is considered to be beyond the scope of this book.

The key point illustrated by the shape conversion matrix in Fig. 2.10 is that it is not symmetric: the complementary off-diagonal terms are not the same because of the path dependency. The conditions for converting a square to a parallelogram are not the same as those for converting a parallelogram to a square.

This path dependency and its influence on the symmetry of the matrix is also illustrated in Fig. 2.11. This Figure contains a matrix of probabilities relating to parameter state changes. It is assumed here that there are three possible states: A, B or C. The parameter could be in state A, B or C and its probability of changing from one state to another is clear from the subscripts. P_{AA} is the probability that if the parameter is in state A it will stay in state A. P_{AB} is the probability that the parameter will change from state A to state B. P_{AC} is the probability that the parameter will change from state A to state C. Naturally, there are three associated probabilities when the parameter is in state B and three more when the parameter is in state C. The nine probabilities are listed out in the transition probability matrix shown. This type of matrix is used to study how a parameter might change its state, the state changes being known as a Markov chain. A row of probabilities in the matrix represents the parameter already being in a certain state and then either staying in that state or changing to the two other states — so these three probabilities must add up to unity.

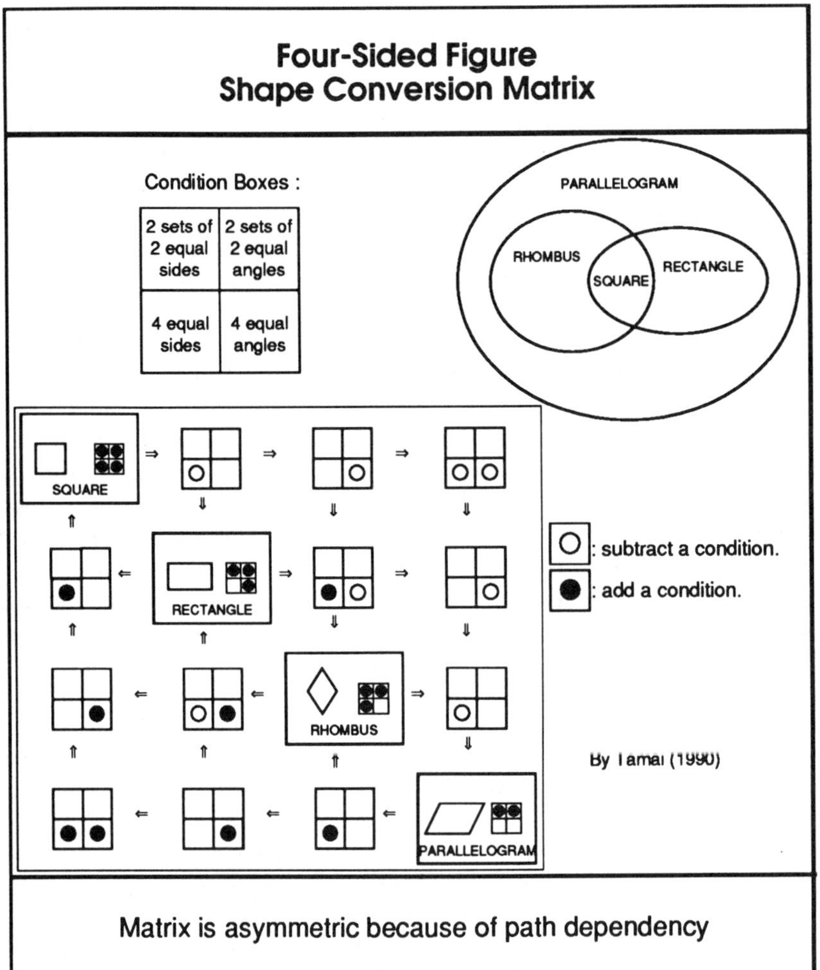

Fig. 2.10 Asymmetry induced by the path dependency of four-sided shape conversion conditions expressed in matrix form.

The reason for presenting this matrix is that it is quite clearly asymmetric. There is no reason why, in general, P_{AB} should be equal to P_{BA} — because of the path dependency. We would not expect the probability of a parameter in state A changing to state B to be the same as the probability of a parameter in state B changing to

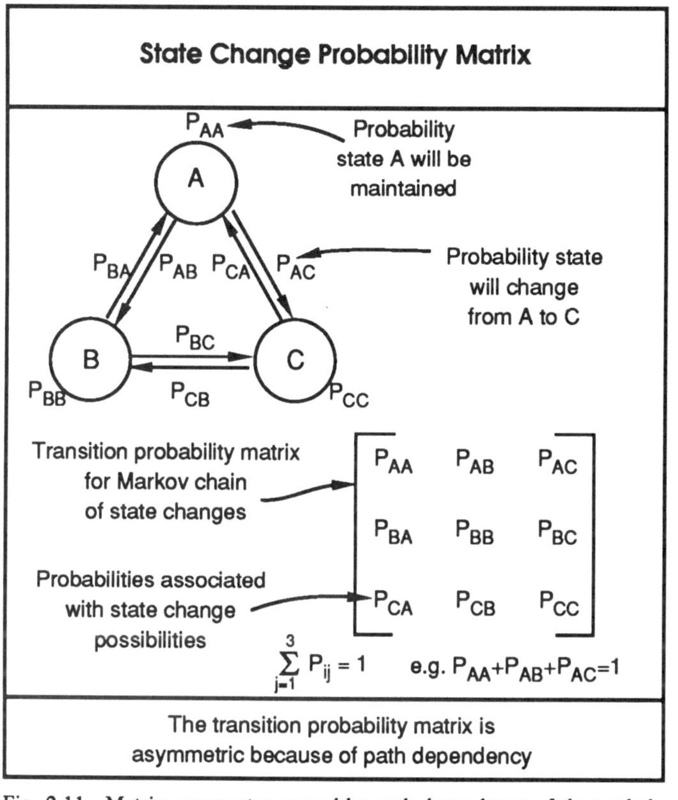

Fig. 2.11 Matrix asymmetry caused by path dependency of the probabilistic state changes being considered.

state A. Thus, again, we are drawn inexorably to the observation that the matrix asymmetry is associated with the existence of path dependency. This is highly likely to be the case for most rock mechanics and rock engineering applications.

The final illustration of the relation between matrix asymmetry and path dependency is shown Fig. 2.12, giving the matrix representation of the alteration in the x-y co-ordinate values of a point when the axes are rotated. In the top left of Fig. 2.12, the rotation of the axes through angle θ is illustrated. In the top right of the Figure, the new co-ordinate values are given with reference to the position of the new axes. These equations are clear by consideration of the components shown in the left part of the Figure. If, now, we detach the x and y coefficients in the equations to form a matrix, the architecture of the matrix can be interpreted in the light of the interaction considerations. The two leading diagonal terms are the same: cosθ. The reason for this is that the main operation, a rotation of the axes, is the same for the x- and y-axis. The off-diagonal terms represent a 'correction factor' that has to be applied because of the interaction between the x-axis and the y-axis resulting from the rotation. This is clear from the geometry of the Figure

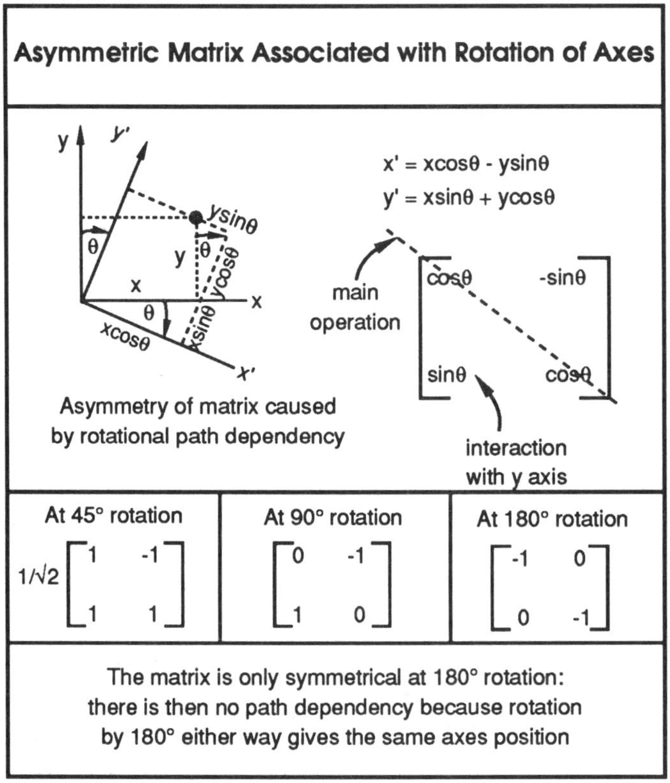

Fig. 2.12 Rotation of axes formula, illustrating that the associated matrix is asymmetric — except for the one case of rotation by 180°.

and the location of the individual components. However, the two interaction terms, –sinθ and sinθ, are not the same because of the minus sign. (If the rotation were to be anti-clockwise rather than clockwise, the two interaction terms would be interchanged.) **Thus, the asymmetry of the matrix is caused by the rotational path dependency.**

The three lower matrices shown in Fig. 2.12 are also very interesting to interpret. In the first case, when there is a 45° rotation, the leading diagonal terms and the off-diagonal terms have the same magnitude, apart from the reversal of the sign in the off-diagonal terms. This is because at 45° degrees the main operation and the interactive operation are the same. In the second case, with a 90° rotation, the x and y axes are interchanged. The main operation along the leading diagonal becomes zero and the interactive terms simply serve to switch the axes, with the direction accounted for by the sign of the interaction terms. Finally, in the third case with a 180° rotation, the axes remain the same: it is simply the positive and negative directions which are being reversed. The leading diagonal terms reverse the signs of the x and y values, and the off-diagonal terms are zero — because at this rotation value there is no interaction between the axes.

This last rotational case refered to, with a 180° rotation of the axes, is the only case where the matrix is symmetrical. This is because there is no path dependency: it does not matter if the axes are rotated clockwise through 180° or anti-clockwise through 180°, they arrive at the same position; hence there is no rotational path dependency and no matrix asymmetry.

This example completes the discussion on matrix symmetry and asymmetry. Because rock engineering mechanisms are almost always path dependent, the interaction matrices that will be studied in this book will be asymmetric.

2.3 MATRIX RESOLUTION

We noted in the previous Section that the interaction matrix could have any number of leading diagonal terms, from two onwards. It was noted that with two leading

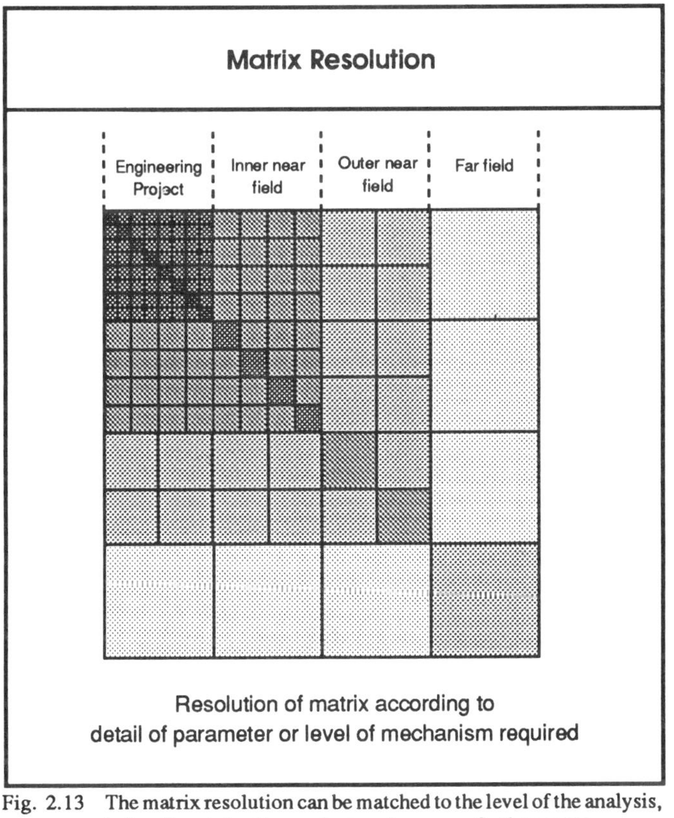

Fig. 2.13 The matrix resolution can be matched to the level of the analysis, being finest for the project and coarser further away.

diagonal terms there are two interaction terms, with three leading diagonal terms there are six interaction terms and with twelve leading diagonal terms there are one hundred and thirty-two interaction terms. In general, for a matrix with N leading diagonal terms, there are $N^2-N = N(N-1)$ interaction terms.

The subject of matrix resolution must now be addressed. The term 'matrix resolution' refers basically to how many leading diagonal terms are present: the larger the number of leading diagonal terms, the finer the resolution. This in turn will be related to the objective of the analysis and hence the reason for compiling

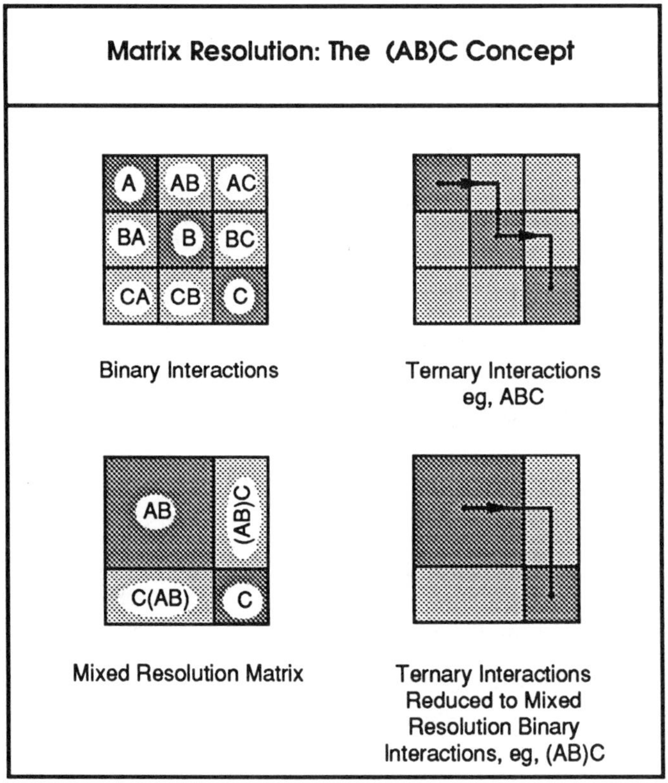

Fig. 2.14 Matrix resolution and the ability to represent multiple interactions as binary interactions in a 'mixed resolution' matrix.

the matrix. Therefore, we must invoke the 'top-down' overview perspective because, so far, the matrix has been described in the synthetic sense (cf. Icon 1 at the beginning of Chapter 1 and Fig. 1.5). We now need to consider the total system that the matrix represents and utilize the analytic (as opposed to synthetic) structure of the matrix. This 'top-down' division is illustrated in Fig 2.13 in which the engineering project, the inner 'near field', the outer 'near field' and the 'far field' are represented.

The engineering project sub-matrix will contain those leading diagonal terms directly relevant to the project. The near field refers to the zone of rock which is not part of the engineering project but which is affected by the engineering. The far field is the region of rock not affected by engineering. (The term 'not affected' is used in an engineering sense, i.e. there is no *significant* effect — as opposed to, for example, elasticity theory where any change theoretically causes a disturbance for an infinite distance).

It can be seen in Fig. 2.13 that the matrix can be resolved according to the complexity of analysis required. But what level of matrix resolution is actually required for a specific engineering project? And how are 'multiple mechanisms' represented, rather than just the binary ones represented by the off-diagonal terms of the interaction matrix described so far?

It is convenient at this stage to introduce a symbolic method of indicating matrix resolution when two or more parameters may have been combined. This is shown in Fig. 2.14 which illustrates the algebraic method, termed the *(AB)C* concept. The *(AB)* component represents the combination of *A* and *B* into one matrix element. It was noted earlier that the basic interaction matrix represents binary interactions, i.e. each off-diagonal term represents a mechanism combining two of the parameters, with the independent variable determined by the clockwise rotation. Thus, the basic interaction matrix with leading diagonal subjects *A* and *B* represents the binary

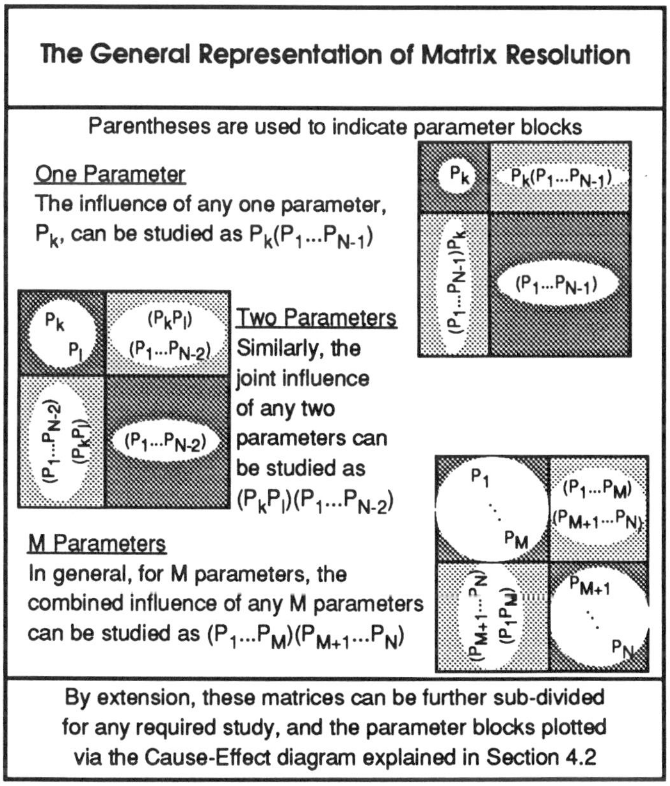

Fig. 2.15 Binary interactions between composite parameter blocks as a general representation of matrix resolution.

interactions *AB* and *BA*. If, now, we wish to consider the combination of three parameters, such as *A* and *B* and *C*, this is termed a ternary interaction. The word 'ternary', meaning composed of three variables, is the mathematical word for a three-variable interaction, matching the word 'binary' for two variables.

Thus, the ternary and, indeed, higher level interactions can be established as either a pathway through the matrix of binary interactions (the upper part of Fig. 2.14) or as a binary interaction where two of the parameters have been combined

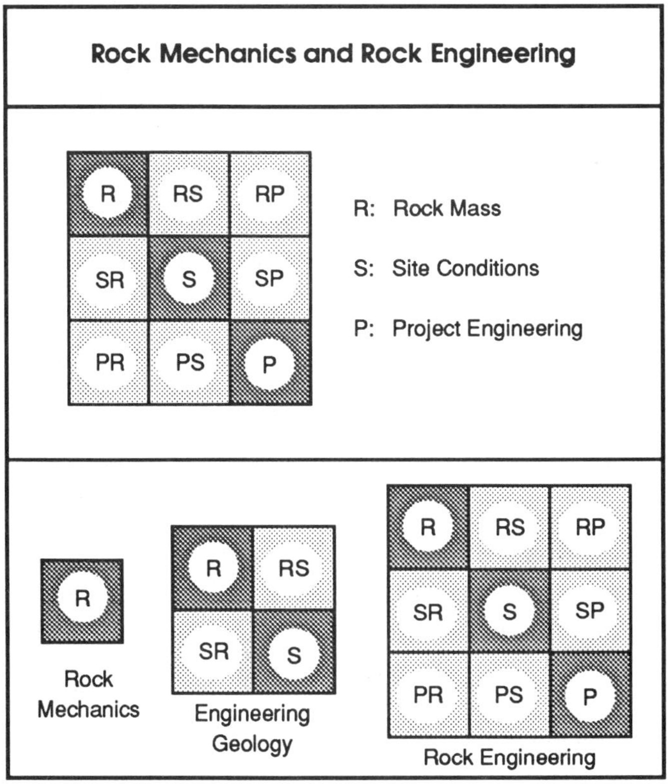

Fig. 2.16 The coarsest possible matrix, incorporating the rock site and project components, in a rock engineering analysis.

(the lower part of Fig. 2.14 for two of the ternary interactions). Naturally, we not only have to consider the *combination* of variables *A*, *B* and *C*, but also the *permutation*: *ABC* is only one of the six permutations of *A*, *B* and *C*. In the lower part of Fig. 2.14, the two ternary interactions represented are *(AB)C* and *C(AB)* because it is the *A* and *B* element that have been combined.

In Fig. 2.15, this concept is extended to the general representation of matrix resolution to show that, with the parentheses being used to indicate parameter blocks, the influence of one parameter, two parameters or *M* parameters can be studied within the context of an *NxN* matrix. (Note that, the use of *n* and *N* are interchangeable in this book: *n* refers to a general matrix with dimension *n*; whereas,

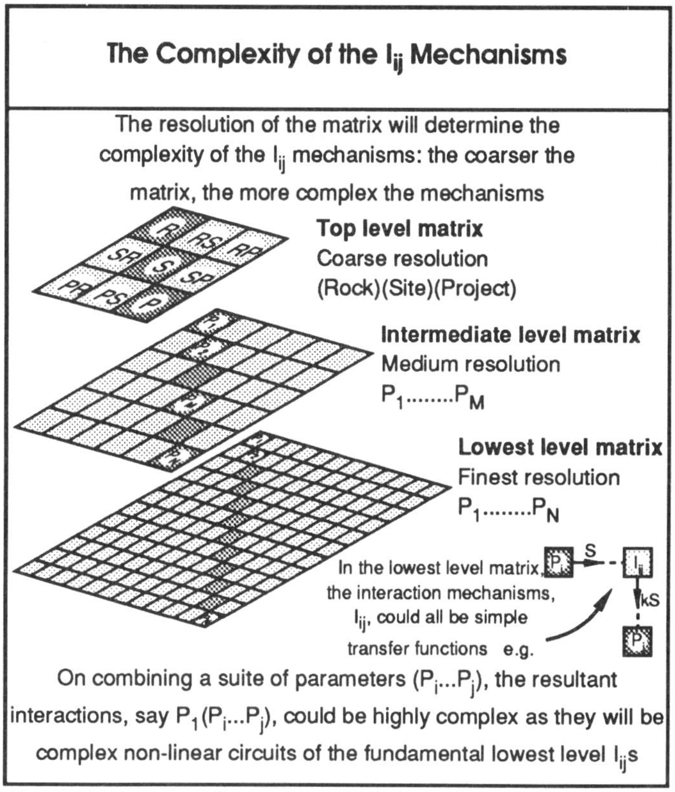

Fig. 2.17 Coarse, medium and fine resolution matrices for studying rock engineering projects via the parameters and mechanisms.

N refers to a specific matrix with dimension N. However, because this distinction cannot always be unambiguously applied, the n and N terms can be interchanged in any of the analysis in this book — they both refer to the matrix dimension.)

The general concept of matrix resolution leads naturally to consideration of the complexity of the mechanisms in the off-diagonal boxes of the interaction matrix. The simplest interaction matrix that incorporates the distinction between rock, site and project for rock engineering is shown at the top of Fig. 2.16. We simply have R, S and P leading diagonal elements, representing the rock, site and project respectively. R refers to the rock mass mechanisms. S refers to the geomorphological and hydrological conditions. P refers to the project engineering and particularly to the project objectives. Obviously, consideration of the off diagonal terms in this coarsest resolution matrix will be conceptual and will be more related to whether we are dealing with rock mechanics, engineering geology, rock engineering, etc. As intimated in the lower part of Fig. 2.16, the whole 3x3 R, S, P matrix corresponds to nine different facets of rock engineering.

Starting with this R, S, P matrix, the resolution can gradually be increased as shown in Fig. 2.17, where coarse, medium and fine resolution matrices are shown. The intermediate level matrix with medium resolution, leading diagonal parameters

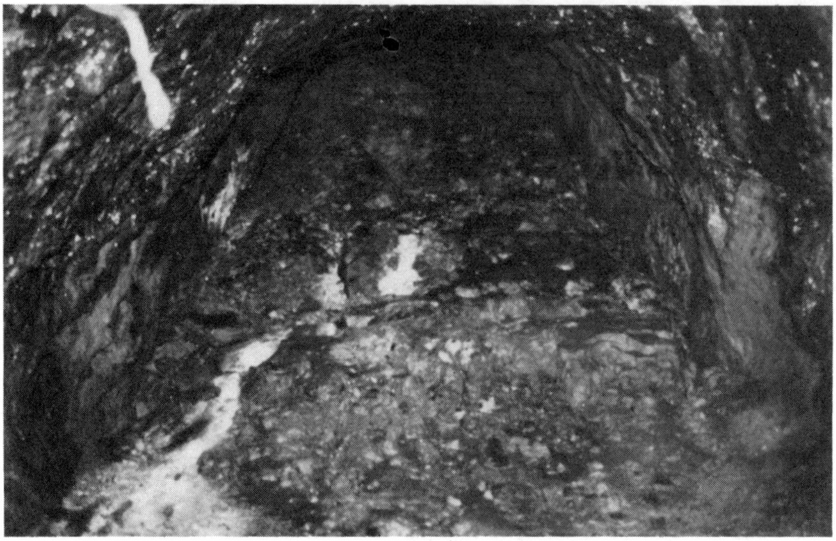

Fig. 2.18 Collapse in a mine development tunnel (kaolin slurry inrush) induced by mine flooding and subsequent pumping out.

$P_I ... P_M$ will be a level where subjects begin to be isolated along the leading diagonal. In the next Chapter, an "Atlas of Rock Engineering Mechanisms" for slopes and underground excavations is presented. These are matrices with twelve leading diagonal terms and with sketches of the mechanisms that can occur in the off-diagonal boxes. Even though the matrix dimension of 12 immediately generates 132 off-diagonal mechanisms for consideration, it will be seen from the Atlas that, depending on the two leading diagonal terms in question, the mechanism may still be extremely complex.

It is only by descending to the lowest level matrix with the finest resolution that we can isolate the fundamental mechanisms shown in Fig. 2.17 as P_I–P_N. In fact in Fig. 2.17, the lowest level matrix interactions could all be equivalent to simple transfer functions, and hence be readily identifiable, measurable and characterizable. The author has not established the working dimension of this 'lowest level' matrix, but it will be in the order of 100x100 or higher. (There are obvious practical limitations to the dimension of the mechanisms matrix, but here we are considering the theory; later, we will discuss the ramifications of the theory and the optimal matrix level at which to establish engineering procedures.)

Naturally, as we ascend away from this finest resolution matrix to coarser resolutions, the mechanisms represented by the off-diagonal boxes can easily become highly complex, as they will represent complex non-linear circuits of the fundamental lowest level interactions.

Consider the photograph in Fig. 2.18. This illustrates a collapse in a mine near Cayeli in Turkey on the Black Sea coast. This collapse occurred after the mine had been flooded and the water subsequently pumped out. How complex is this collapse mechanism? The inrush, shown in a development drive, is a whitish kaolin-type natural slurry. Is this an interaction between the water and the rock in the lowest level matrix in Fig. 2.17 or is it at a higher level, where the mechanisms are more complex and parametric path dependency may be involved? The purpose of showing the photograph in Fig. 2.18 is to illustrate the 'systems thinking' that develops from this approach.

Consider, alternatively, the photograph of a natural fracture in a rock core shown in Fig. 2.19. This is the 300th of just over 1000 discontinuities studied by the author in a 400 m length of granodiorite core. Note the very clear evidence of slickensiding on the core surface which results from earlier movement on this pre-existing fracture. How do we characterize this fracture? How many parameters are involved? What level of matrix resolution is being illustrated here?

Fig. 2.19 Slickensided discontinuity surface in a granodiorite rock core.

Clearly, there are major implications for characterization procedures and testing methods inherent in Fig. 2.17 and consideration of the questions above. Naturally, in developing a characterization procedure, it is helpful to have previously decided to which level of matrix resolution the characterization procedure applies. It is unnecessary to point out the hazards of characterizing the rock at different levels of mechanism resolution.

In Fig. 2.20, these ideas are re-expressed in the form of cascading matrices. Essentially the process of increasing the matrix resolution means that there is generally a sub-matrix within each element of the higher level matrix. Hence, as the matrix resolution is increased, there is an associated parameter hierarchy of the leading

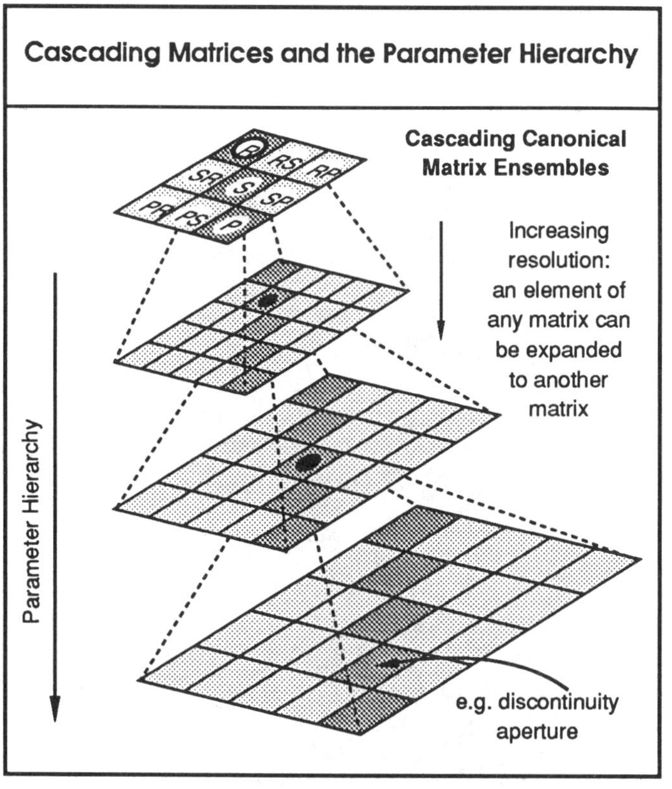

Fig. 2.20 Increasing matrix resolution considered as cascading canonical matrix ensembles with an associated parameter hierarchy.

diagonal terms and a suite of cascading canonical matrix ensembles — until finally, at the lowest level, we hope to reach some fundamental, identifiable and measurable quantity. 'Discontinuity aperture' is shown as the example in Fig. 2.20: whether this is at the lowest level or not depends on the degree to which we wish to pursue the analysis. If the parallel plate model is used to represent discontinuity aperture, then, indeed, the discontinuity aperture is at the fundamental base level matrix resolution: it is one scalar value. If, however, we wish to represent the discontinuity aperture as it really is (cf. the discontinuity surface shown in Fig. 2.19), then we still have a long way to descend!

As yet, there is no list of rock properties and no list of rock engineering mechanisms, so we will have to create these as the systems methodology continues

Fig. 2.21 Pre-split rock face, A9 road near Perth, Scotland.

to be developed. There are many ramifications implicit in Fig. 2.20, not least of which is that we actually need to know the potential leading diagonal and off-diagonal terms. Without these, we do not know which are the relevant parameters, nor the potential mechanisms that could influence the rock engineering.

To end this Chapter on the principles of interaction matrices and matrix resolution, the reader might like to ponder on the photograph in Fig. 2.21 which shows a pre-split rock face of a cutting on the A9 road near Perth in Scotland. What are the interactions between the *R*, *S* and *P* components here? How has the rock affected the blasting? How has the blasting affected the rock? How has the drilling affected the blasting? How has the rock affected the drilling etc? At what matrix resolution level are the mechanisms? At what matrix resolution should we study this total system? What was the engineering objective? Has the engineering objective been achieved?

3

Atlas of Rock Engineering Mechanisms

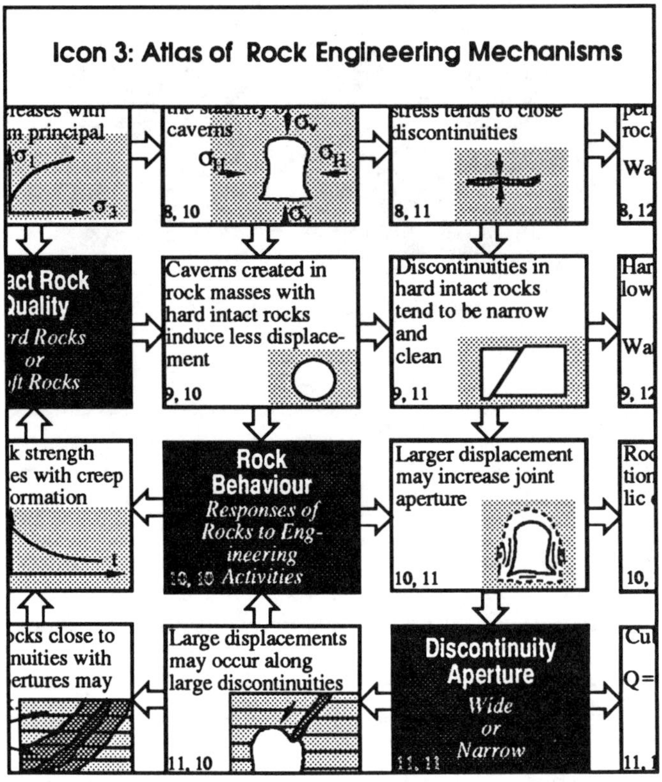

By choosing appropriate leading diagonal parameters for given engineering circumstances and then identifying the interaction components in the off-diagonal boxes, an atlas of rock engineering mechanisms can be created.

In this Chapter, the interaction matrix device described in Chapter 2 will be used to generate and illustrate generic 12x12 matrices for slopes and underground excavations. Firstly, appropriate leading diagonal terms are chosen for each subject and then the mechanisms in the off-diagonal terms are identified.

The 4x4 interaction matrix in Figure 2.7 contains the four leading diagonal terms: *rock structure, rock stress, water flow and construction.* As has been discussed, matrices with such coarse resolution will, by definition, contain integrated information; to explore a subject in more detail, it is necessary to increase the size of the matrix — and to choose appropriate leading diagonal terms. Naturally, these leading diagonal terms will be dictated by the rock, site and project. However, the leading diagonal terms chosen here are for generic matrices to demonstrate the method and to provide two working matrices which can be used to illustrate the matrix coding systems which will be discussed later.

It is important to note that the analytic method is still being used. The matrix is not compiled by taking an existing mechanism and finding out where it is best located in the matrix. The matrix is compiled by considering, with reference to each off-diagonal box, the influence of one leading diagonal parameter on another: we may or may not be able to fill the box with known information. Once the matrix has been generated, it provides a checklist of relevant rock engineering mechanisms and their locations within the matrix. Each matrix can be thought of as a map; together, these matrix maps form an atlas of rock engineering mechanisms.

3.1 ROCK SLOPES

The twelve leading diagonal terms chosen for the generic slopes matrix are

1. Overall Environment	*Geology, climate, seismic risk, etc.*
2. Intact Rock Quality	*Strong, weak, weathering susceptibility*
3. Discontinuity Geometry	*Sets, orientations, apertures, roughness*
4. Discontinuity Properties	*Stiffness, cohesion, friction*
5. Rock Mass Properties	*Deformability, strength, failure*
6.* In Situ *Rock Stress	*Principal stress magnitudes/directions*
7. Hydraulic Conditions	*Permeability, etc.*
8. Slope Orientation etc.	*Dip, dip direction, location*
9. Slope Dimensions	*Bench height/width & overall slope*
10. Proximate Engineering	*Adjacent blasting, etc.*
11. Support/Maintenance	*Bolts, cables, grouting, etc.*
12. Construction	*Excavation method, sequencing, etc.*

Naturally, we might wish to alter these for a specific slope location or requirement. The purpose here is to illustrate how the use of the interaction matrix, with these twelve leading diagonal terms, enables the identification of 132 off-diagonal terms representing the associated rock engineering mechanisms. These are illustrated in the four quadrants of the matrix shown in Figures 3.1a to 3.1d.

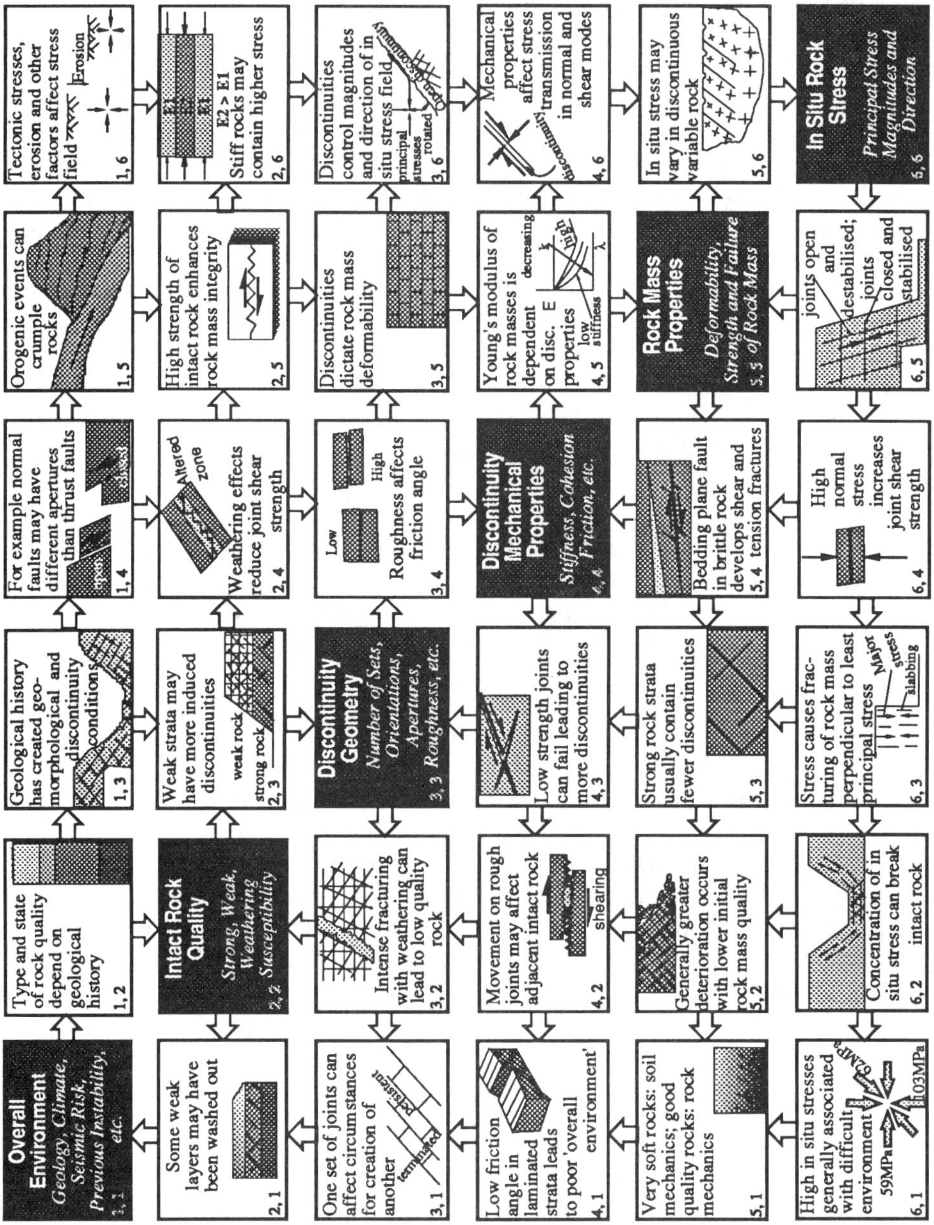

Fig. 3.1a Top left quadrant of 12X12 generic slopes matrix. Matrix leading diagonal terms are

1. Overall Environment
2. Intact Rock Quality
3. Discontinuity Geometry
4. Discontinuity Mechanical Properties
5. Rock Mass Properties
6. *In Situ* Rock Stress
7. Hydraulic Conditions
8. Slope Orientation and Location
9. Slope Dimensions
10. Proximate Engineering
11. Support/Maintenance
12. Construction

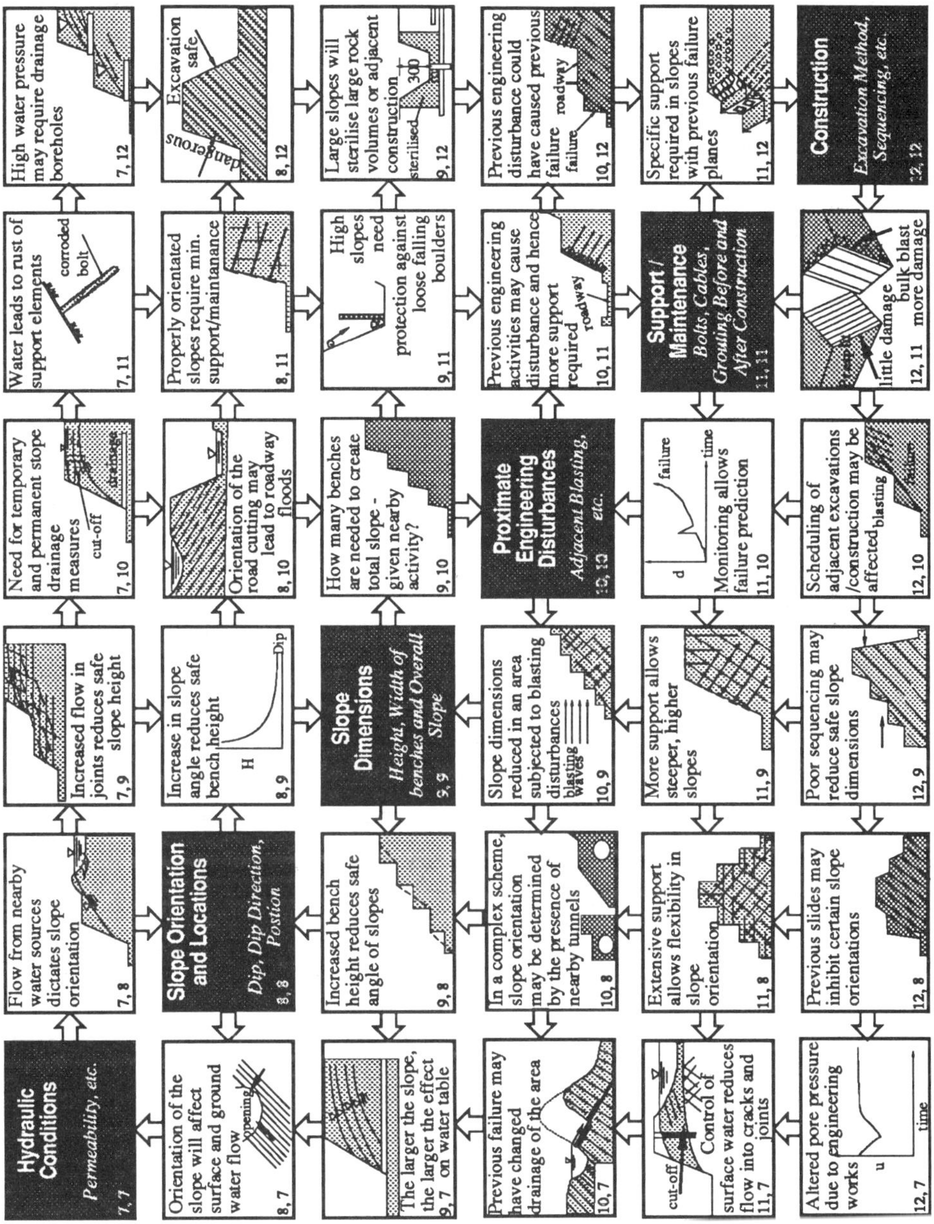

Fig. 3.1b Bottom right quadrant of 12X12 generic slopes matrix. Matrix leading diagonal terms are

1. Overall Environment
2. Intact Rock Quality
3. Discontinuity Geometry
4. Discontinuity Mechanical Properties
5. Rock Mass Properties
6. *In Situ* Rock Stress
7. Hydraulic Conditions
8. Slope Orientation and Location
9. Slope Dimensions
10. Proximate Engineering
11. Support/Maintenance
12. Construction

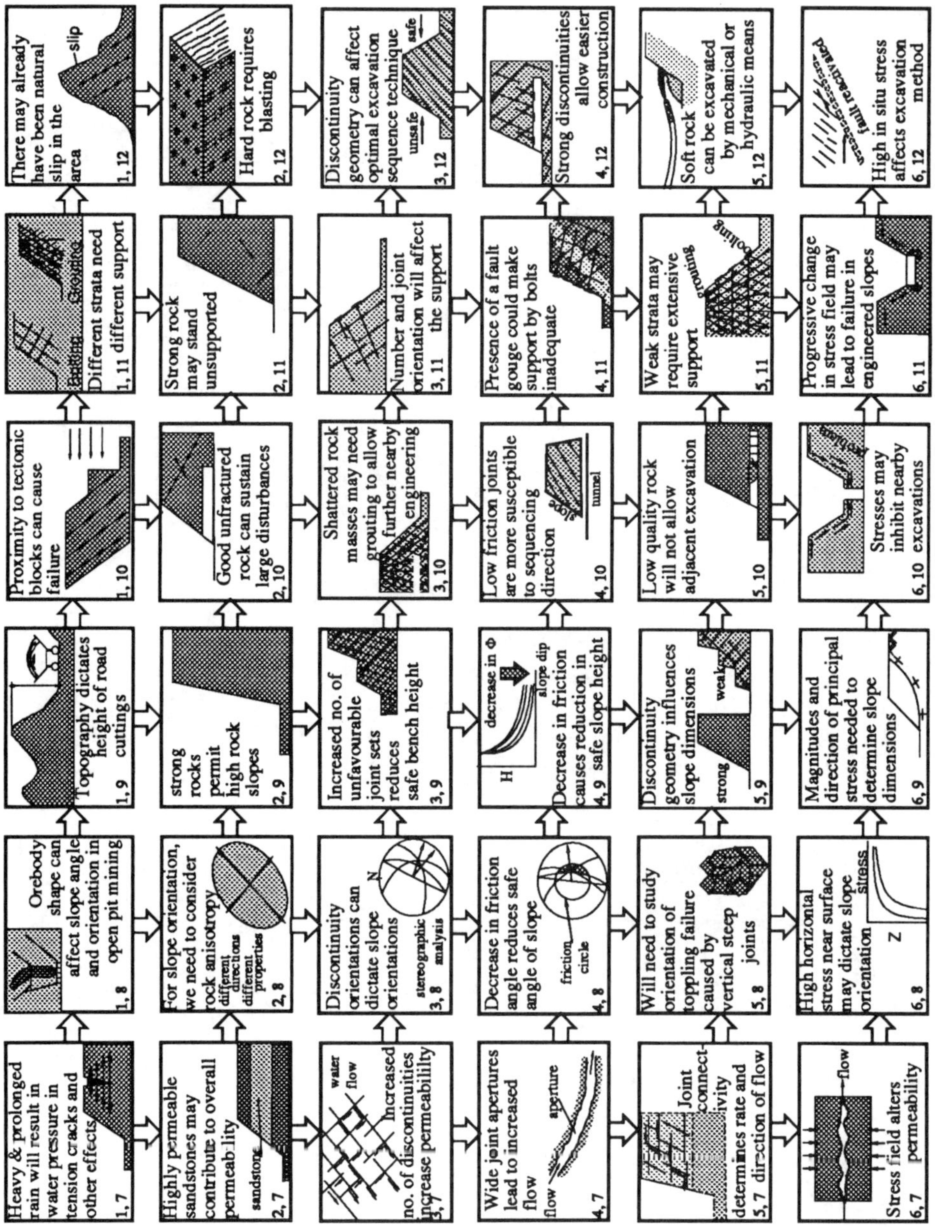

Fig. 3.1c Top right quadrant of 12x12 generic slopes matrix. Matrix leading diagonal terms are

1. Overall Environment
2. Intact Rock Quality
3. Discontinuity Geometry
4. Discontinuity Mechanical Properties
5. Rock Mass Properties
6. *In Situ* Rock Stress
7. Hydraulic Conditions
8. Slope Orientation and Location
9. Slope Dimensions
10. Proximate Engineering
11. Support/Maintenance
12. Construction

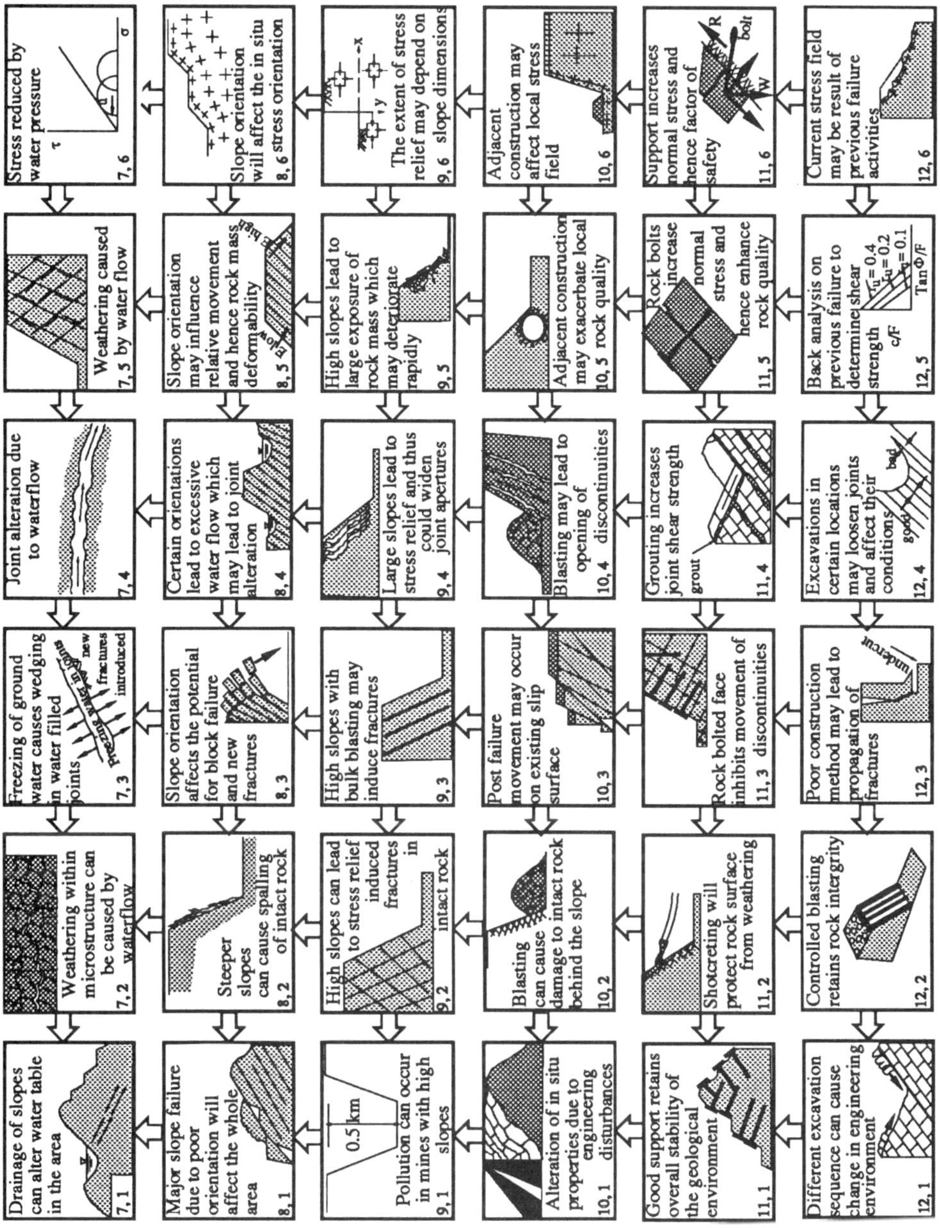

Fig. 3.1d Bottom left quadrant of 12x12 generic slopes matrix. Matrix leading diagonal terms are

1. Overall Environment
2. Intact Rock Quality
3. Discontinuity Geometry
4. Discontinuity Mechanical Properties
5. Rock Mass Properties
6. *In Situ* Rock Stress
7. Hydraulic Conditions
8. Slope Orientation and Location
9. Slope Dimensions
10. Proximate Engineering
11. Support/Maintenance
12. Construction

3.2 UNDERGROUND EXCAVATIONS IN ROCK

The twelve leading diagonal parameters for a generic underground excavations matrix are given below. These are similar to those of the generic slopes matrix but the emphasis is somewhat different to reflect the different circumstances.

*1. **Excavation Dimensions***	*Excavation size and geometry*
*2. **Rock Support***	*Rockbolts, concrete liner, etc.*
*3. **Depth of Excavations***	*Deep or shallow*
*4. **Excavation Methods***	*Tunnel boring machines, blasting*
*5. **Rock Mass Quality***	*Poor, fair, good*
*6. **Discontinuity Geometry***	*Sets, orientations, distributions, etc.*
*7. **Rock Mass Structure***	*Intact rock and discontinuities*
8. **In Situ** ***Rock Stress***	*Principal stress magnitudes and directions*
*9. **Intact Rock Quality***	*Hard rocks or soft rocks*
*10. **Rock Behaviour***	*Responses of rocks to engineering activities*
*11. **Discontinuity Aperture***	*Wide or narrow*
*12. **Hydraulic Conditions***	*Permeabilities, water tables, etc.*

Again, these twelve leading diagonal terms generate 132 off-diagonal interactions or mechanisms, which are illustrated in Figures 3.2a to 3.2d overleaf.

The reader is encouraged to study the off-diagonal boxes in Figures 3.1 and 3.2 and to note where existing information has been identified and where our knowledge is lacking. For example, in Figure 3.2b, the Box 12,8 represents the influence of Hydraulic Conditions on *In Situ* Rock Stress — the well-known effective stress law. However, in Figure 3.2d the Box 12,6 represents the influence of Hydraulic Conditions on Discontinuity Geometry, "Water flow may silt up discontinuities altering their effective distributions", a subject about which we know very little.

For those Figures indexed c) and d) in which the sub-matrix contains only off-diagonal components, (where the leading diagonal boxes are absent), reference to the listings of the leading diagonal terms in the Figure caption will assist in locating any particular Box ij, simply remembering that Box ij represents the influence of leading diagonal term Box i on leading diagonal term Box j.

Study of the choice of leading diagonal terms and the off-diagonal terms that are consequentially invoked for each of these main matrices illustrated produces a whole variety of questions. Should the leading diagonal terms all be of the same 'quality', perhaps all expressible in energy terms? Knowing that the information in each off-diagonal box is illustrative only and not necessarily comprehensive, how could we ensure that all rock engineering mechanisms are explicitly identified? Obviously, it does not matter for the interaction matrix concept in what order the leading diagonal terms are arranged, but would some convention help? How are the leading diagonal terms to be chosen knowing the rock engineering objective? How can the most important leading diagonal terms be identified? Can any of the information in the matrix be ignored? In the next Chapter, we will concentrate on developing a method for establishing parameter interactive intensity and dominance in order to provide an approach to these questions.

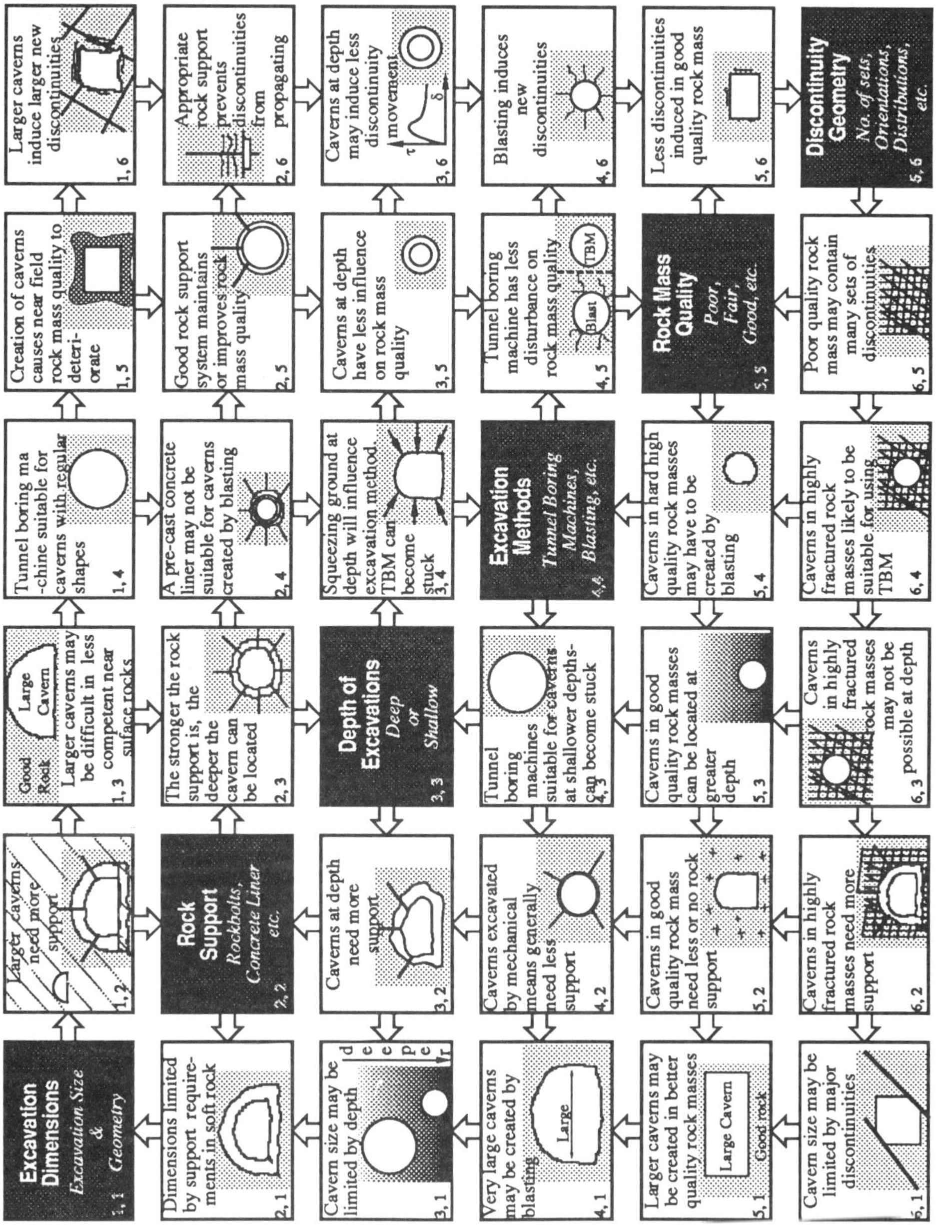

Fig. 3.2a Top left quadrant of 12x12 generic underground excavations matrix. Leading diagonal terms are

1. Excavation Dimensions
2. Rock Support
3. Depth of Excavations
4. Excavation Methods
5. Rock Mass Quality
6. Discontinuity Geometry
7. Rock Mass Structure
8. *In Situ* Rock Stress
9. Intact Rock Quality
10. Rock Behaviour
11. Discontinuity Aperture
12. Hydraulic Conditions

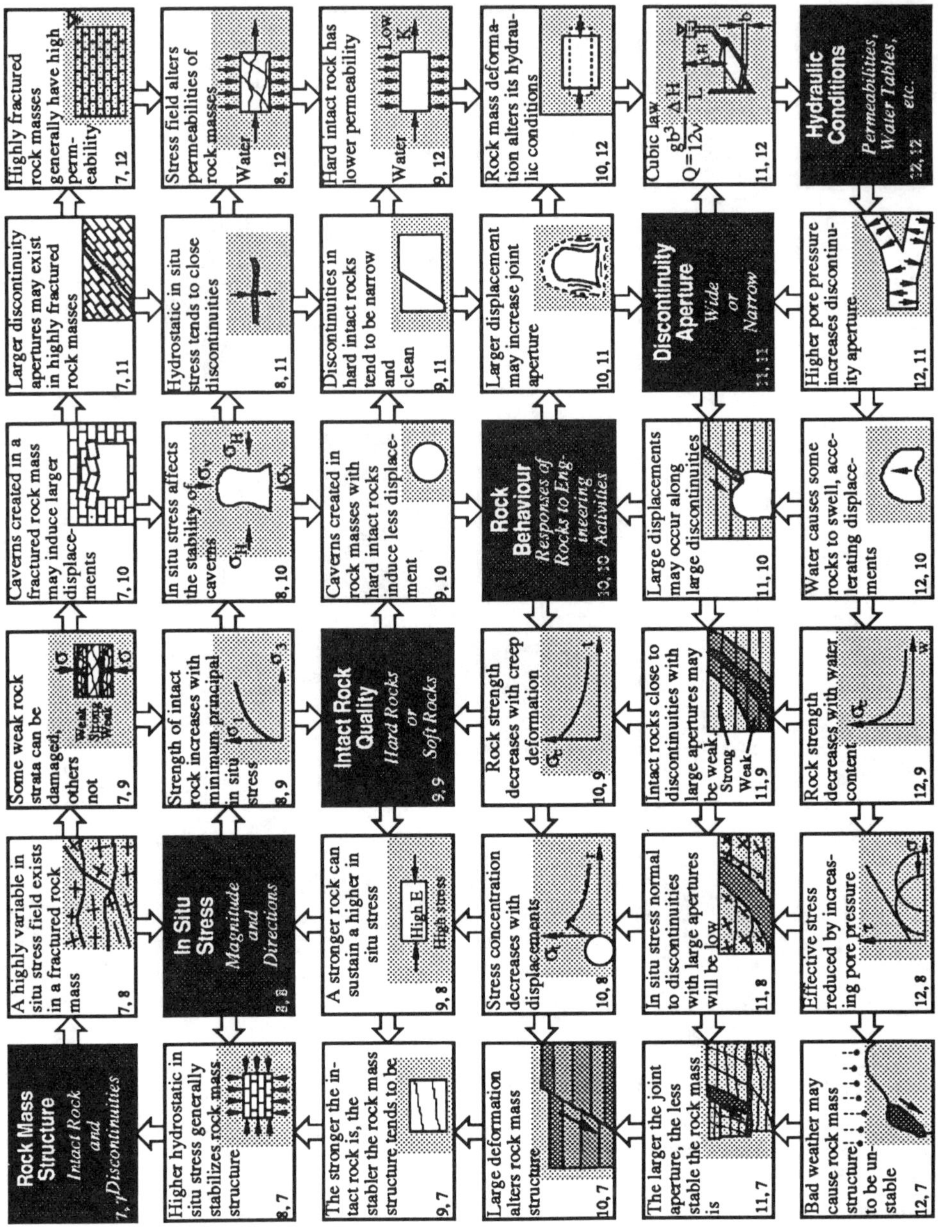

Fig. 3.2b Bottom right quadrant of 12x12 generic underground excavations matrix. Leading diagonal terms are

1. Excavation Dimensions
2. Rock Support
3. Depth of Excavations
4. Excavation Methods
5. Rock Mass Quality
6. Discontinuity Geometry
7. Rock Mass Structure
8. *In Situ* Rock Stress
9. Intact Rock Quality
10. Rock Behaviour
11. Discontinuity Aperture
12. Hydraulic Conditions

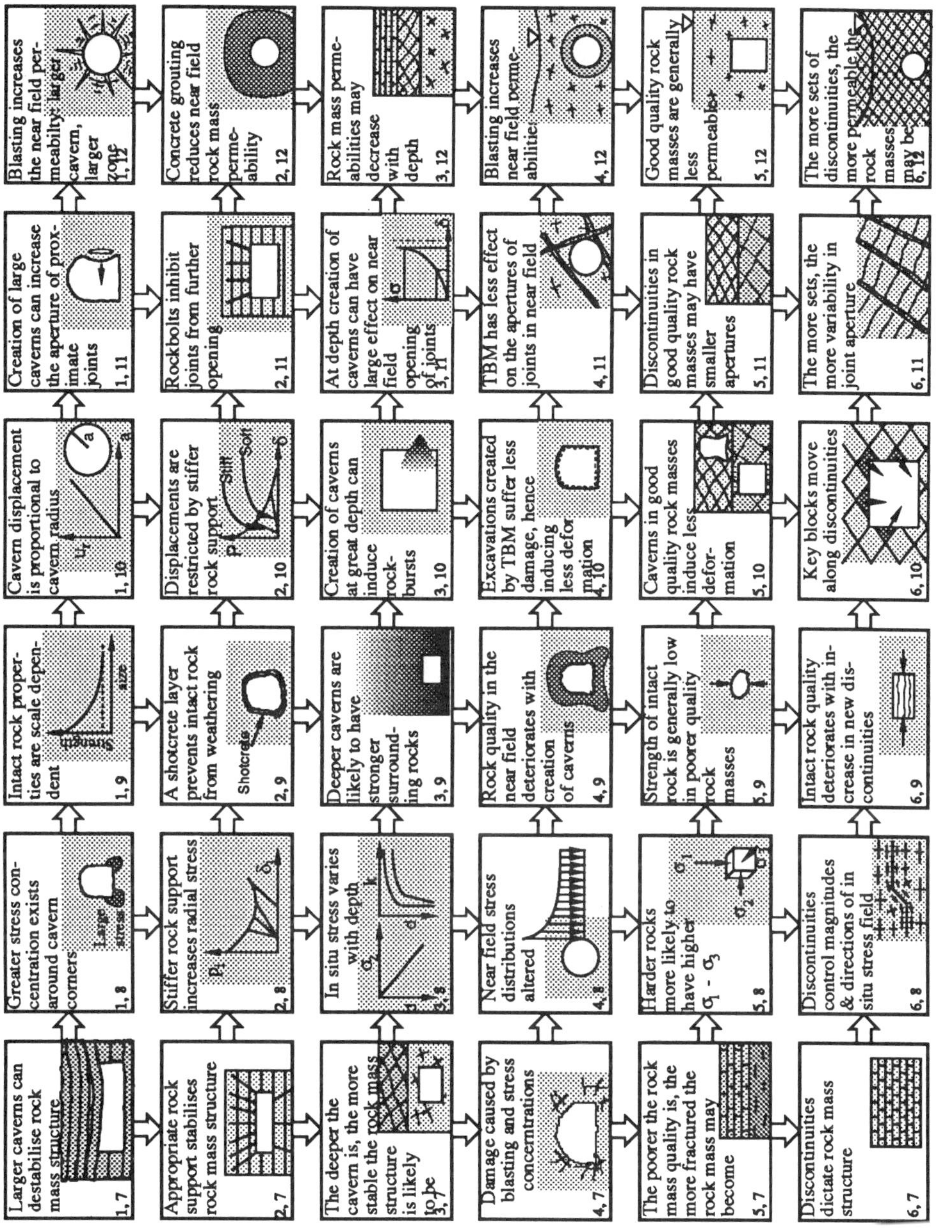

Fig. 3.2c Top right quadrant of 12x12 generic underground excavations matrix. Leading diagonal terms are

1. Excavation Dimensions
2. Rock Support
3. Depth of Excavations
4. Excavation Methods
5. Rock Mass Quality
6. Discontinuity Geometry
7. Rock Mass Structure
8. *In Situ* Rock Stress
9. Intact Rock Quality
10. Rock Behaviour
11. Discontinuity Aperture
12. Hydraulic Conditions

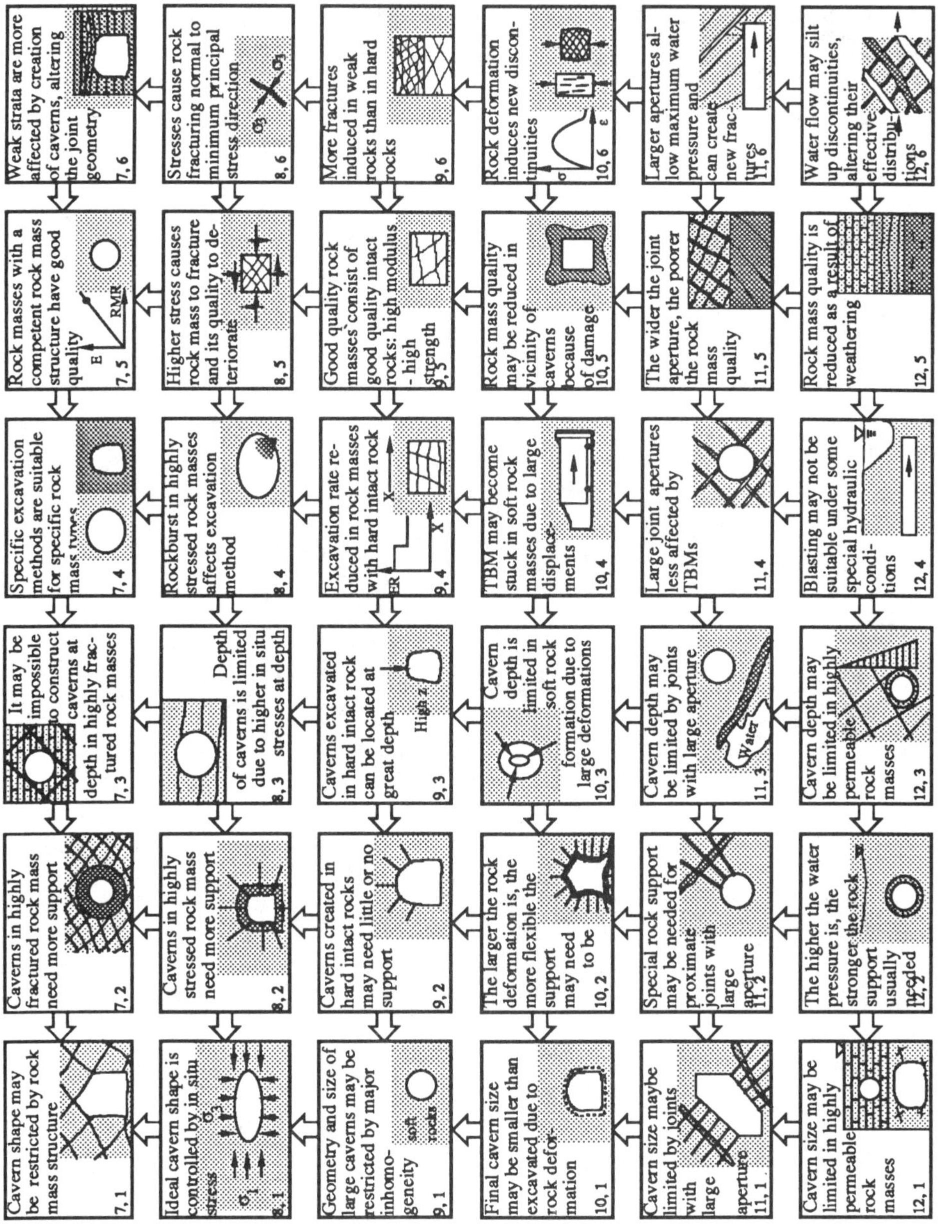

Fig. 3.2d Bottom left quadrant of 12x12 generic underground excavations matrix. Leading diagonal terms are

1. Excavation Dimensions
2. Rock Support
3. Depth of Excavations
4. Excavation Methods
5. Rock Mass Quality
6. Discontinuity Geometry
7. Rock Mass Structure
8. *In Situ* Rock Stress
9. Intact Rock Quality
10. Rock Behaviour
11. Discontinuity Aperture
12. Hydraulic Conditions

4

Parameter Interaction Intensity and Dominance

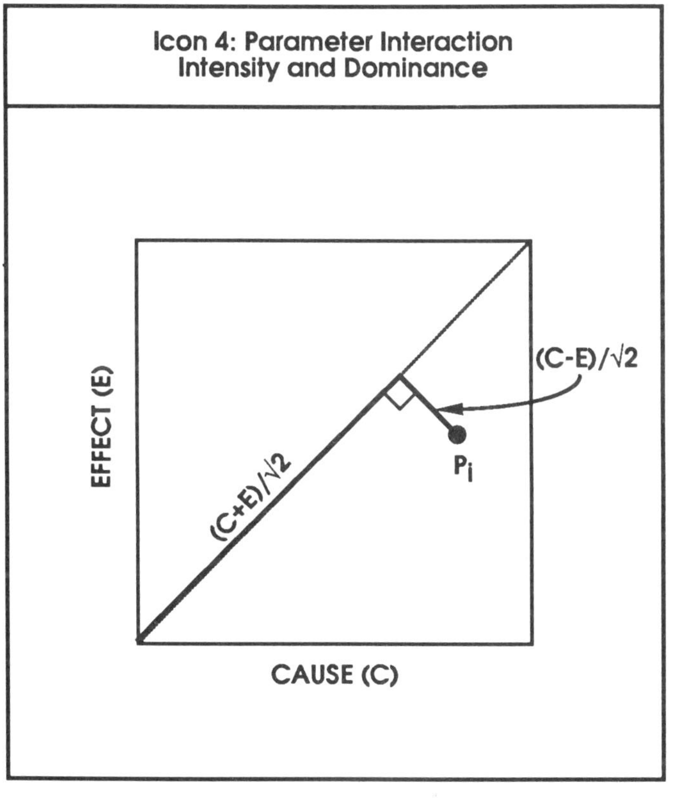

By coding the interaction matrix components and then summing the values in the row and column through each parameter, 'cause' and 'effect' co-ordinates are generated, indicating a parameter's interaction intensity & dominance.

Following the explanation of the interaction matrix device in Chapter 2 and the Atlas of Rock Engineering Mechanisms presented in the last Chapter, we can now consider how to quantify parameter significance, through the two measures of parameter interaction intensity and dominance. (These values are illustrated in Icon 4 on the previous page as the distance along the diagonal and the perpendicular distance from this diagonal to the parameter point.) The precursor to the analysis

Fig. 4.1 Quarry bench clearly illustrating the effects of the rock discontinuities on blasting.

must be a method of coding the interaction matrix in order to establish how the parameters influence each other via the mechanisms.

We know that some parameters will have a greater effect on the rock engineering system than others and, similarly in turn, the system will have a greater effect on some parameters than others. In Fig. 4.1, there is a photograph of a quarry bench. It is evident from this photograph that any engineering activity in the rock will be dominated by the master joints, which are clearly visible in the photograph. Not only will they affect construction, but in they are affected by construction. For example, dilation induced by blasting is evident in the photograph. A further example is shown in Fig. 4.2 in which part of a tin mine is shown. The stope

Fig. 4.2 Stope in a tin mine in Cornwall, England, illustrating the strong influence of the rock structure on stope geometry.

geometry is a direct reflection of the discontinuities in the rock. However, these are qualitative and subjective observations. Some form of quantitative translation of such observations and our engineering knowledge of the mechanisms is required. This is achieved by coding the interaction matrices and studying the interaction intensity and dominance of each parameter.

4.1 MATRIX CODING

In Fig. 4.3, five matrix coding methods are shown. These cover the entire spectrum from a simple decision as to whether a particular mechanism is operating or not to a complete numerical analysis of the mechanism.

The first method is the binary approach. The mechanism is either considered to be switched on or off: if it is switched on, we code the off-diagonal component with a value of unity; if it is switched off, we code the box with a value of zero. In the atlas in Chapter 3 containing 12x12 matrices, there are 132 off-diagonal boxes in each case. Even a simple count of the mechanisms which are operating using the binary coding method would be helpful as a start. For example, if for a particular 12x12 matrix we obtained nine off-diagonal boxes with a value of one, and a hundred and twenty-three boxes with a value of zero, we would know that the system is not very interactive. Conversely, if the number of 'switched on' boxes is over, say, sixty-six, then the matrix will no longer be sparse and the system will be significantly interactive.

The second coding method shown in Fig. 4.3 is the 'expert semi-quantitative' (ESQ) method. This is really an extension of the binary method except that there are five categories into which the mechanism can be placed, ranging from zero to four, corresponding to 'no', 'weak', 'medium', 'strong' and 'critical' interactions respectively. Through this type of coding, there is greater sensitivity than the simple on/off switch of the binary method. It may appear at first sight that it might be difficult to allocate the mechanism to a particular category, but the reader will find that having undertaken the exercise once or twice, the categorization is usually definite. Moreover, as we will see later, the procedure does allow for some variability in the category allocation. As a consequence of coding the matrix by this method, we have a potential maximum sum of five hundred and twenty-eight when adding all the box values in the matrix. However, we will not be considering the total of all the matrix values directly, but rather the sums of the rows and columns through each of the leading diagonal terms.

The third method is the slope of the P_i *vs.* P_j plot. The two graphs in Fig. 4.3 illustrate this method. If the graph of P_j as a function of P_i is a horizontal line, P_j does not depend on P_i. Alternatively, if there is some approximately linear relation,

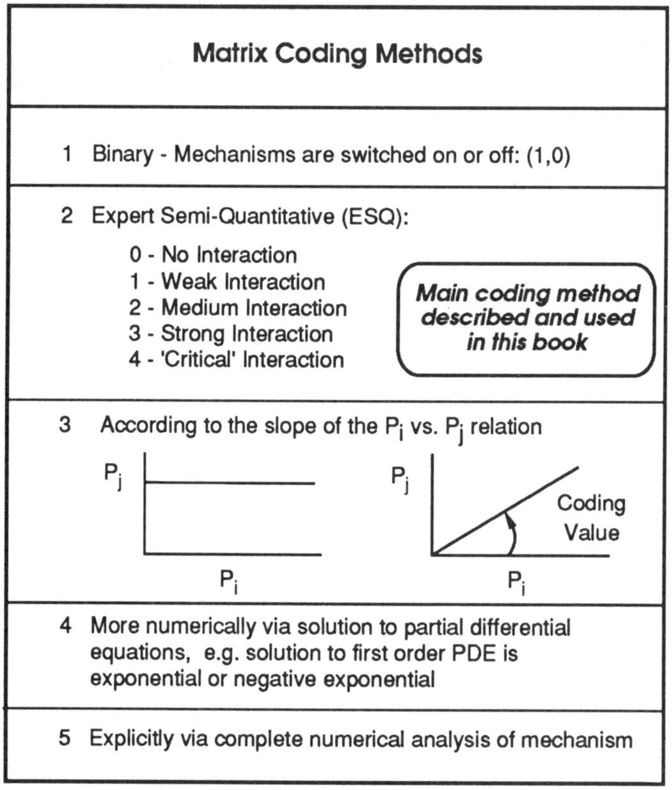

Fig. 4.3 The five basic coding methods from a simple on–off coding to complete numerical analysis of the mechanism.

then we can code the mechanism box by the slope of the line, which indicates the rapidity by which P_j changes as a function of P_i. In order to utilize this method comprehensively throughout a matrix, we would have to know the P_i *vs.* P_j curves for all the off-diagonal boxes. Also, there would be the problem that in many cases the relation would no doubt be non-linear. Thus, although this is a more 'scientific' extension of the first two methods, we are unlikely to have sufficient information to utilize the method. However, for certain sub-matrices for which more knowledge is available, it might well be possible to evaluate the links between some parameters by this method.

The fourth method is to adopt a direct systems approach. We assume that all the mechanisms in the boxes can be represented by partial differential equations and estimate the constants associated with the solutions. For example if we assume that all the mechanisms can be represented by first order partial differential equations, then exponential or negative exponential functions characterize the P_i *vs.* P_j relations. Despite the unknowns associated with this approach, it might be considered preferable to method three because the exponential approximation is likely to be an improved description. Of course, if we assume that second order partial differential equations apply, the solutions will be even more realistic — but it is a daunting task to try to estimate the 132 approximations for each of the off-diagonal boxes in a 12x12 matrix. However, there is considerable potential in this approach.

The fifth method is the explicit method, which assumes that there is enough knowledge of the mechanisms to support numerical analysis, and hence to be able to code each of the boxes coherently according to the actual mechanics. This information will definitely not be available for the complete matrix but, since we wish to make use of all available information, this method could perhaps be linked to the second and fourth methods in due course. At that level of analysis, it will be essential for the leading diagonal parameters to have the same 'quality', i.e. to be expressed in the same, or compatible, units.

For the presentations in this book, we will use the second method shown in Fig. 4.3 (the ESQ method) with the allocation of the mechanisms being to one of five categories. This ESQ coding method is viable for any matrix and will serve to demonstrate how the systems approach is developed and hence how in principle other coding methods could be applied.

4.2 THE CAUSE–EFFECT PLOT

Consider now the diagram in Fig. 4.4 which shows the generation of the cause and effect co-ordinates. The main parameters, P_i, are listed along the leading diagonal with 'construction' as the last box. We interpret the meaning of the rows and columns of the matrix, as highlighted in this diagram by the row and column through P_i. From the construction of the matrix, it is clear that the *row* passing through P_i represents the influence of P_i on all the other parameters in the system. Conversely, the *column* through P_i represents the influence of the other parameters, i.e. the rest of the system, on P_i. Once the matrix has been numerically coded, we can find the sum of each row and each column. If now, we think of the influence of P_i on the system, we can term the sum of the row values as the 'cause' and the sum of the column terms as the 'effect', designated as co-ordinates (C,E). Thus, C represents the way in which P_i affects the system; and E represents the effect that the system has on P_i. Note that construction itself has (C,E) co-ordinates, repesenting the post-construction and pre-construction mechanisms respectively.

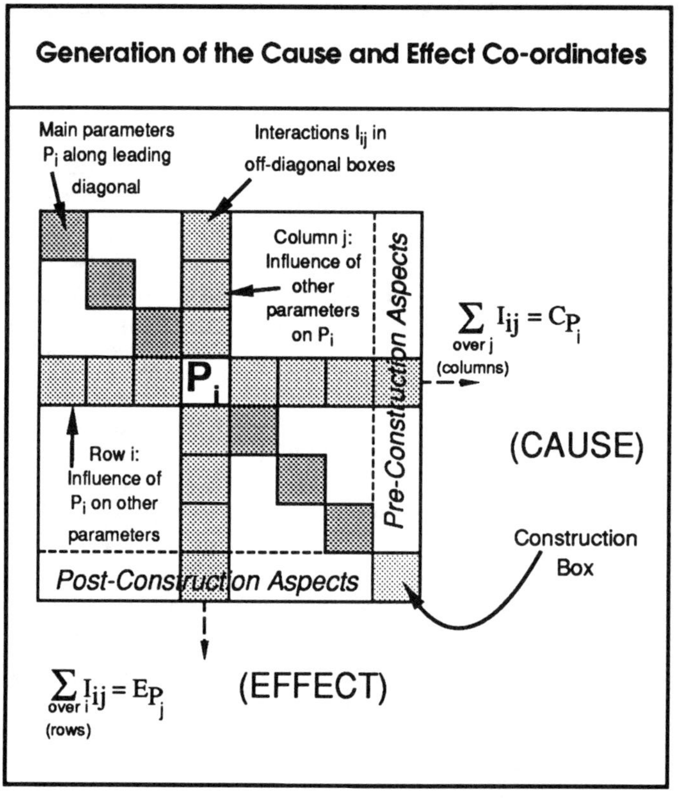

Fig. 4.4 Summation of coding values in the row and column through each parameter to establish the cause and effect co-ordinates.

Utilizing the second method of matrix coding (the ESQ method) and with a 12x12 matrix, each row and column can have a maximum summated value given by 11x4 = 44. The value of the leading diagonal box itself, P_i, has no value. In other words, we assume here that the parameter does not affect itself: the parameter can only affect itself via another parameter and an associated pathway through the matrix, a subject we will come to later.

The co-ordinate values for each parameter can now be plotted in cause and effect space. In order to introduce this idea, we start with the (C,E) plot for two parameters as shown in Fig. 4.5. Here the summation of the rows in the columns is just the single value of the off-diagonal blocks. Thus, the maximum value of the rows and columns is four. Note the mean value of the co-ordinates for P_1 and P_2, which is also plotted on the diagram.

In general, we would not use a 2x2 matrix in isolation, but it might occur as a result of combining parameters using the idea discussed earlier. For example, P_1 could be the effect of water and P_2 could be tunnelling construction. One of the interaction boxes would then be the influence of water on tunnelling (tunnelling is more difficult when water is present) and the other interaction box would then be

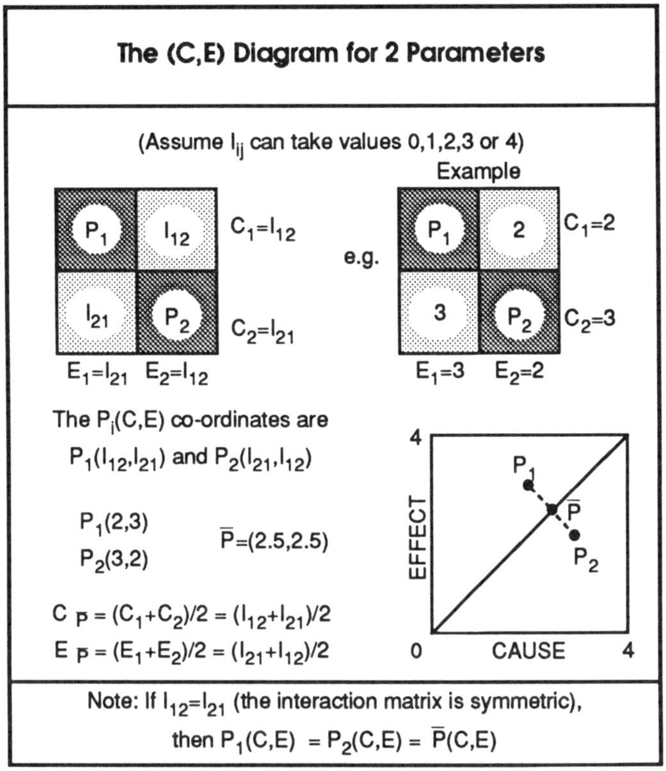

Fig. 4.5 Example cause *vs.* effect plot for two parameters.

Fig. 4.6 Drilling ahead of the tunnelling face to detect the existence of any adverse hydrological conditions.

the influence of tunnelling on the flow of water (water flows into the tunnel). In Fig. 4.6 there is a photograph of drilling through a hole in the tunnelling shield into the rock ahead of the tunnel to detect the presence of any water and hence whether this particular matrix box is likely to be switched on or switched off (cf. the binary coding method in Fig. 4.3).

In Fig. 4.7, the (C,E) diagram is extended to three parameters. There are three cases: a symmetric matrix and two asymmetric matrices. The reader is encouraged to consider these cases carefully to establish how, in the first case, the parameters

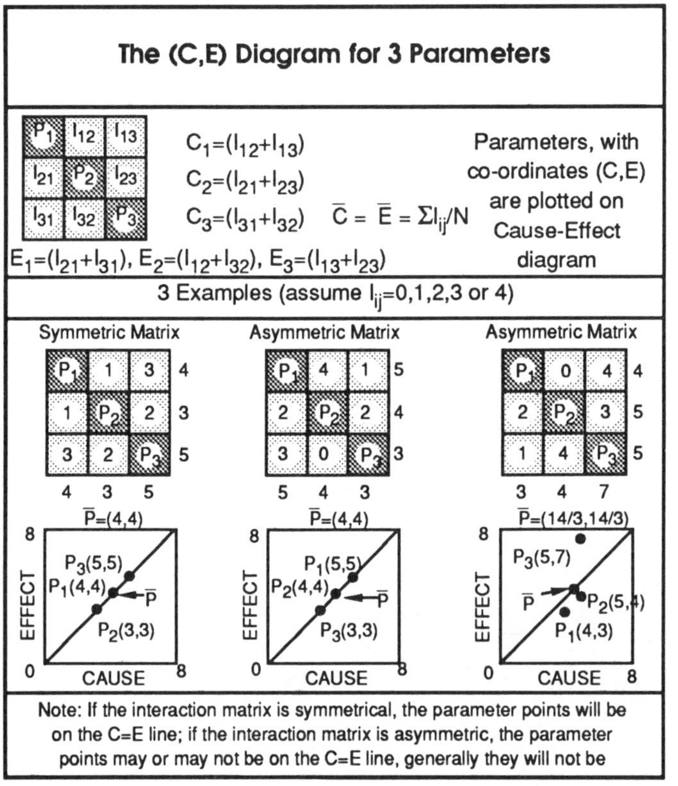

Fig. 4.7 Extension of the cause *vs.* effect plot to three parameters, for both symmetric and asymmetric matrices.

plot on the $C{=}E$ line and how, in the second case, they also plot on this line, despite the fact that the second matrix is asymmetric. In the third case, however, the points do not lie on the $C{=}E$ line for the asymmetric matrix. The interesting case and the one which is generally encountered is the third one, where the points are distributed away from the $C{=}E$ line.

The (C,E) diagram is now extended to N parameters in Fig. 4.8. In this general case, the parameters plot as a cloud in C,E space and might form particular types of constellations for the particular systems that are being represented. Consideration of the position of these parameter points is crucial in the development and utility of the systems theory presented here.

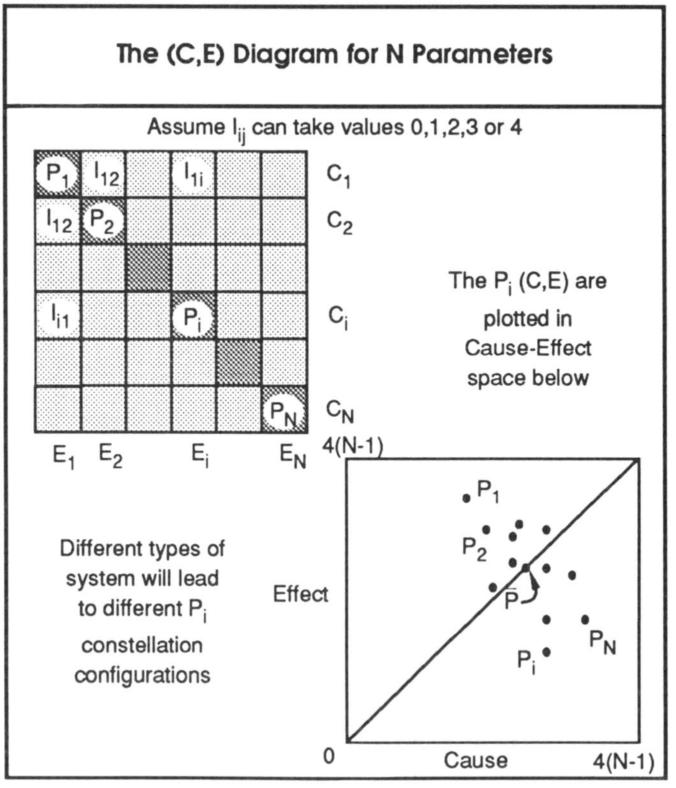

Fig. 4.8 Extension of the cause *vs.* effect plot to N parameters, and the associated constellation of parameter points.

In Fig. 4.9, the generation of a *C,E* plot is explicitly shown. For the matrix element coding values indicated, the row and column sums are given and the parameter points plotted in the lower right hand graph in Fig. 4.9. The position of the points in cause *vs.* effect space indicates the mode of interaction of each parameter. The most interactive parameter is P_4 because it has the highest $C+E$ value. The least interactive parameters are P_1 and P_3 because they have the least $C+E$ value. The most dominant parameter is P_1 because it has the highest $C–E$ value: P_1 affects the system far more than the system affects P_1. The most subordinate parameters are P_2 and P_4 because they have the least $C–E$ value (taking the sign into account). Note that the average of the parameter co-ordinates plots on the $C=E$ line, which follows Theorem 1 to be discussed later.

The mathematics of the coding values and the generation of the *C,E* plots are straightforward but for a moment let us reflect on how these might be applied in practice. In Fig. 4.10, there is a photograph of the cutting head of a five metre diameter tunnel boring machine. For the operation of this machine, one might wish to know which rock parameters and which machine parameters are the most dominant and interactive. Is the compressive strength of the rock or the presence of rock j

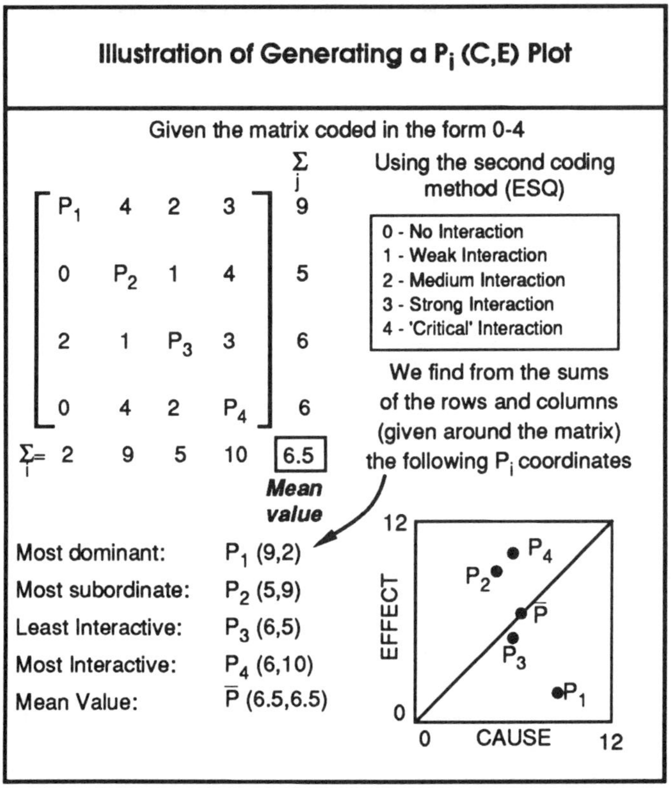

Fig. 4.9 Specific example of a cause *vs.* effect plot for a 4x4 matrix showing different parameter interaction intensity and dominance.

Fig. 4.10 Head of full-face tunnel boring machine illustrating one aspect of rock engineering for which we might wish to establish all the machine–rock interactions.

joints or the presence of water the most dominant? Is the machine thrust, the machine torque or the ability of the side rams to remain in place the most dominant machine parameter? Tunnelling is an example of a system. This is because it involves many interactive parameters. All these parameters and more will contribute to the overall operation. By establishing the matrix dedicated to a particular rock,

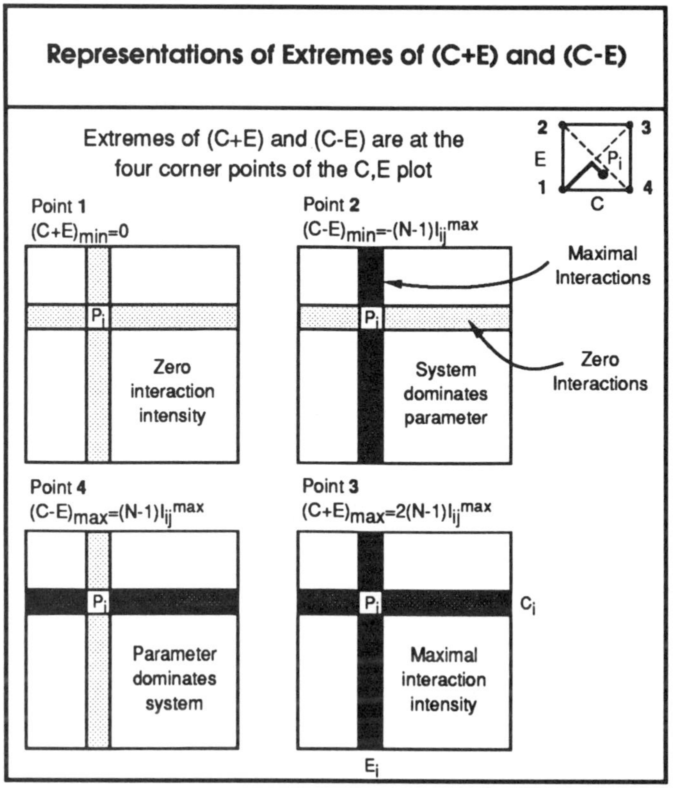

Fig. 4.11 The maximum and minimum values of the parameter interaction intensity and dominance, expressed in matrix form.

site and project, and coding the matrix, we can establish the most interactive and most dominant parameters. Once the information is available, we can consider how to optimize the tunnelling system and how changing one parameter may affect the others.

We may well encounter extreme values of $C+E$ and $C-E$, i.e. extreme values of parameter intensity and dominance. These extreme values are shown in Fig. 4.11 and are manifested at the four corners of the C,E square shown in the Figure.

The reader is referred to the block of four matrices in Fig. 4.11. The case of a parameter having zero intensity will occur when the associated row and column are empty (the top left matrix in Fig. 4.11). The case of maximal parameter interaction intensity occurs when the associated row and column are full and each box has its highest value (the bottom right matrix). The two other extremes, representing maximal parameter dominance and subordinacy, occur respectively when the row is maximized and the column is zero (the bottom left matrix), and when the row is zero and the column is maximized (the top right matrix). Thus, we have the four extreme cases shown in the block of four matrices in Fig. 4.11 of zero interaction intensity, maximal interaction intensity, the parameter dominates the system, and the system dominates the parameter.

4.3 MATRIX SYMMETRY AND THE CAUSE–EFFECT PLOT

It has already been indicated that the symmetry of the matrix plays an important role in reflecting the state of the system: this is very elegantly manifested via the *C,E* plot. If the matrix is symmetric, all the parameter points must lie on the *C=E* line; if the matrix is asymmetric, in general the parameter points will not lie on the *C=E* line. The two cases are illustrated in Figs. 4.12 and 4.13.

When all the parameter points are on the *C=E* line, we only have to consider interactive intensity: the parameters are neither dominant nor subordinate; their effect on the system is the same as the effect of the system on them, because in all cases their *C* and *E* co-ordinates are equal. This is clearly illustrated in Fig. 4.12.

The general case of an asymmetric matrix is illustrated in Fig. 4.13 where the

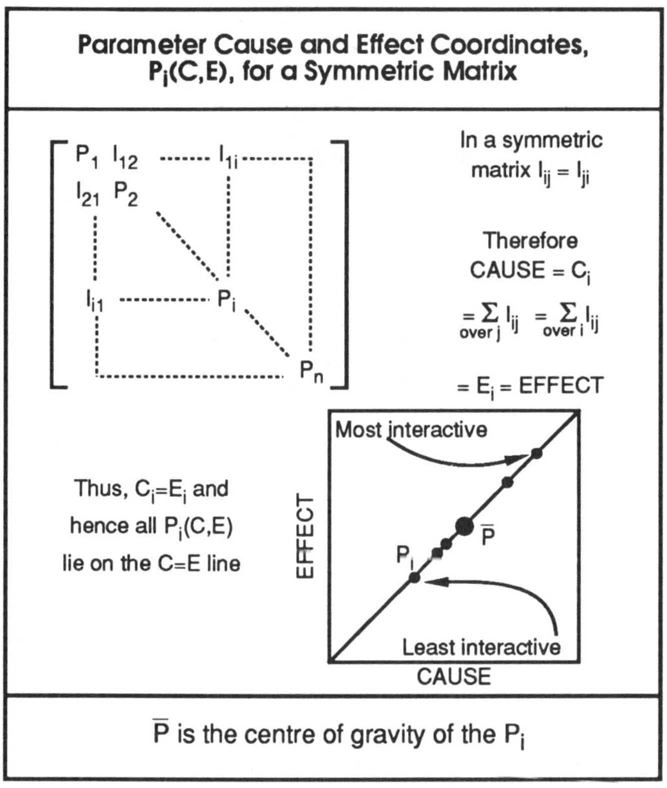

Fig. 4.12 In the cause *vs.* effect plot derived from a symmetric matrix all the parameter points lie on the *C=E* line.

release of the matrix symmetry constraint allows the parameter points the freedom to move away from the $C{=}E$ line. Thus, parameters have the dual characteristics of interactive intensity and dominance. The centre of gravity of the points lies on the $C{=}E$ line because the effect of all the parameters on the system must be the same as the effect of the system on all the parameters (through consideration of which terms are being added in the row and column associated with each parameter).

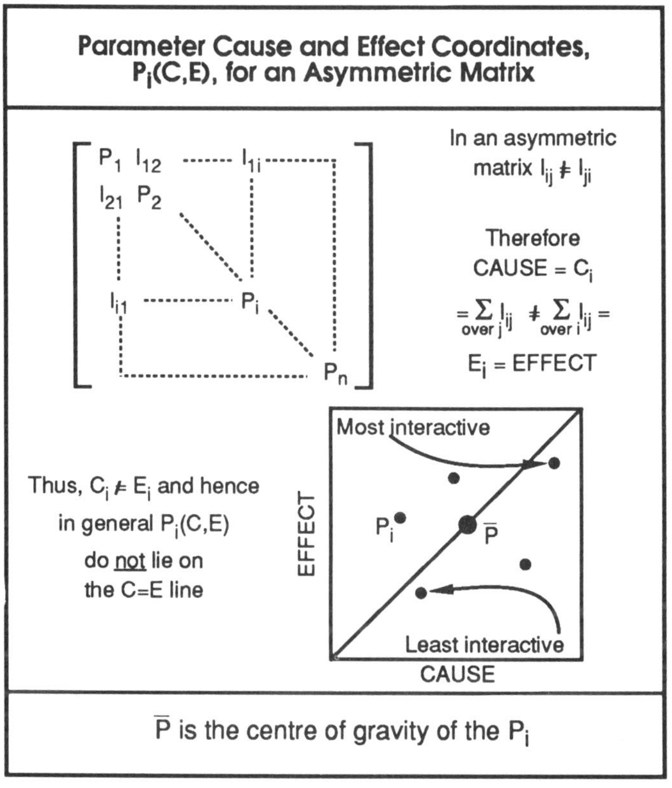

Fig. 4.13 In the cause *vs*. effect plot derived from an asymmetric matrix the parameter points will generally not lie on the $C{=}E$ line.

In Fig. 4.14, a further interpretation is given of the position of a parameter point in the cause *vs*. effect plot. For a particular parameter, the C,E co-ordinates can be considered as the sums of individual co-ordinates formed by the complementary interaction terms in the row and column for each parameter. As indicated in Fig. 4.14, the importance of each of the individual contributions to the total can be assessed visually by this method, with both the intensity and dominance being a cumulative effect of this addition. The parameter point on the cause *vs*. effect diagram is therefore a total vector comprised of the 'complementary pair' component vectors. It is constructive to consider the ways in which the parameter point, with its C,E co-ordinates, can have a high interactive intensity or dominance through the summation of the component vectors.

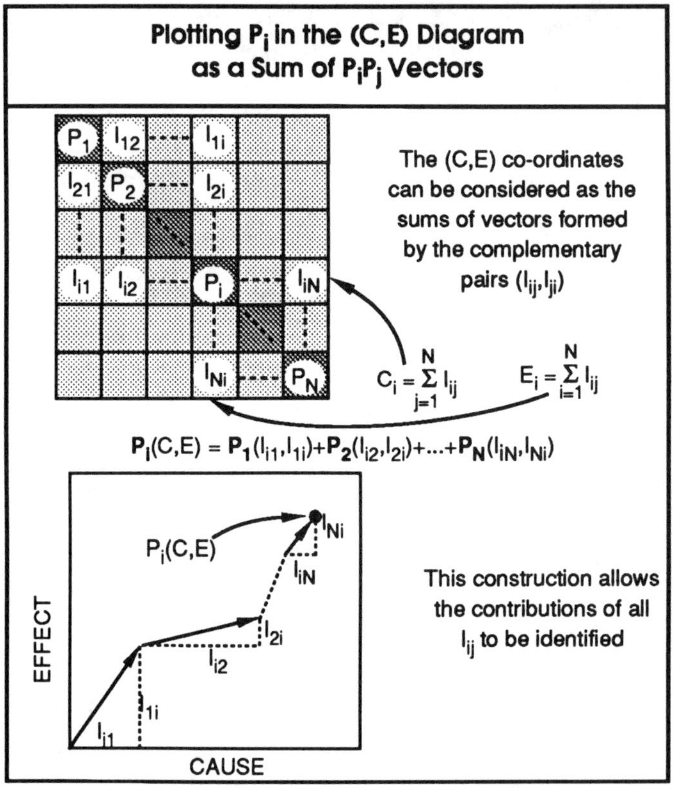

Fig. 4.14 Position of parameter point in cause *vs*. effect space interpreted via the elemental component vectors for each contribution.

4.4 INTERPRETATION OF THE CAUSE–EFFECT PLOT

We are now in a position to refer back to Icon 4 on the first page of this Chapter. The iconic representation is significant because it shows that the parameter intensity can be measured along the $C=E$ line and the parameter dominance can be measured by the perpendicular distance of the parameter point from this line.

The iconic representation is expanded in Fig. 4.15, which illustrates these parameter interaction intensity and parameter dominance characteristics. The two sets of 45° lines in the plot indicate contours of equal value for each of the two characteristics. It is particularly important to note that, whilst the parameter interaction intensity increases monotonically from zero to the maximum, the associated maximum possible parameter dominance value rises from zero to a maximum at 50% parameter interaction intensity and then reduces back to zero at maximum parameter intensity value. The specific numerical values of the two characteristics are $(C+E)/\sqrt{2}$ and $(C-E)/\sqrt{2}$ as indicated in Fig. 4.15.

The way to interpret a C,E plot, therefore, is to assess a parameter's position in the diagram via the two characteristics indicated by the heavy lines in Fig. 4.15.

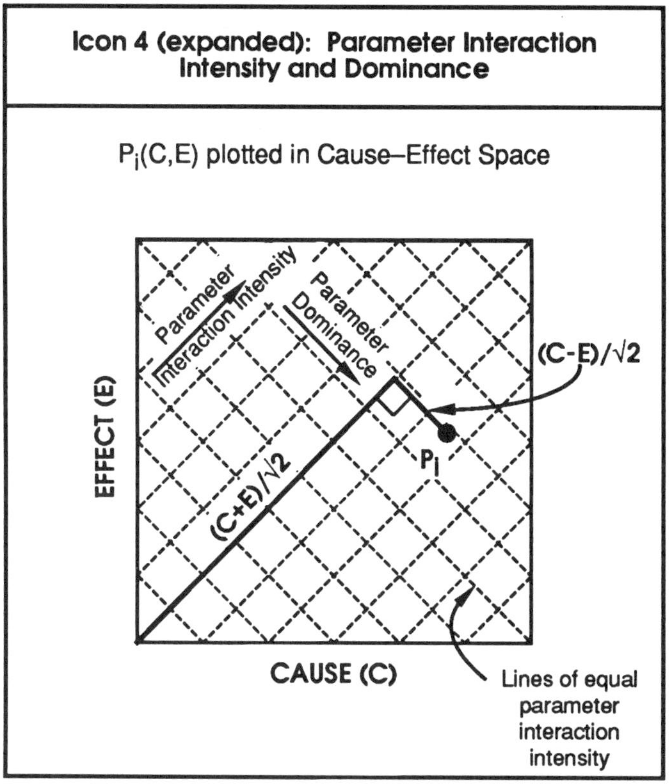

Fig. 4.15 Icon 4 expanded to show lines of equal parameter interaction intensity and dominance.

Fig. 4.16 Discontinuities will be a dominant parameter for most rock engineering: those shown here are in a granitic rock.

Again, the question of how this position is interpreted in practice and where we might expect particular parameters to lie is another fundamental aspect of this systems approach. Consider the discontinuities shown in Fig. 4.16. In any questions of stability of a structure with dimensions at all commensurate with those of the rock shown in the diagram, we would expect the discontinuities to be both interactive and dominant in the system. On the other hand, in Fig. 4.17 there is a photograph of a street in a town in England with no immediate apparent interest. Inspection of the photograph will reveal evidence of tunnelling. The levels of the top of the slats in the fence show an almost perfect inverted normal curve, representing subsidence of the ground surface. The presence of the water on the left-hand side of the photograph indicates that the tunnel is probably being driven perpendicular to the direction of the road. If we included this parameter of subsidence into our matrix analysis of the excavation system, we might expect it to have a negative $C–E$ value, especially if there were no restraints on the subsidence.

Thus, having established the interpretation of the parameter points in C,E space, we should identify the engineering ramifications. Are there particular regions in the C,E plot which have engineering benefit and some which do not? For example, with reference to the subsidence shown in Fig. 4.17 and, if subsidence can occur freely, this parameter will be subordinate to the system: the system will induce subsidence but the subsidence will not affect the operation of the system. However,

Fig. 4.17 Subsidence (note the curve in the top of the fence) which could be a subordinate or a dominant parameter, depending on subsidence constraints (in UK).

if there were planning regulations of any kind which restricted the subsidence to certain levels, the parameter could become dominant as constraints on subsidence would become a major determinant of the system operation. The greater the constraints on subsidence, i.e. the lower the allowable limits for subsidence, the greater will be the dominance of the subsidence in the tunnelling system. Increasingly stringent precautions will have to be introduced to restrict the subsidence levels.

One method of providing engineering guidance directly would be to code the matrix according to whether the mechanism involved had a useful or adverse effect on engineering. In other words, we would code the matrix according to our subjective engineering assessment of the value of the mechanism to our overall objective.

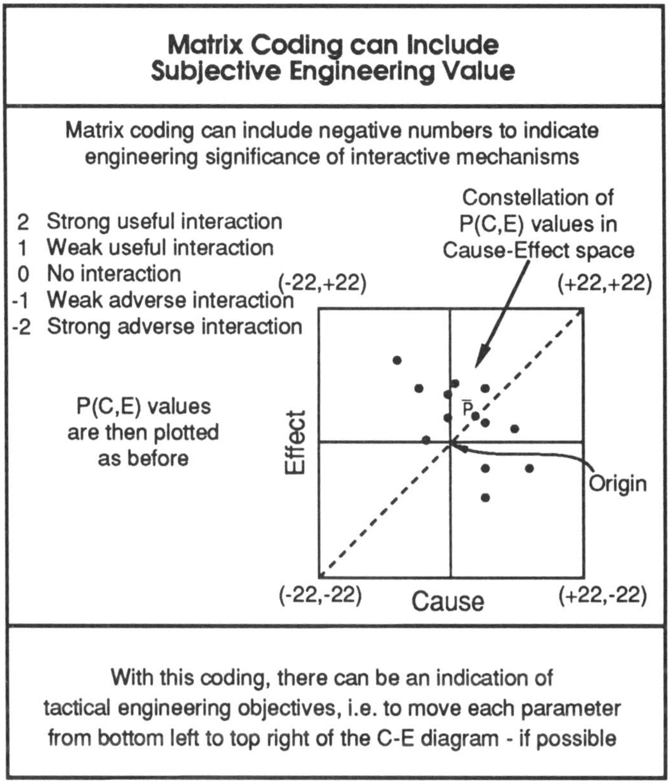

Fig. 4.18 Subjective coding of interaction matrix and associated cause *vs.* effect plot for project-based analysis and interpretation.

Such a subjective coding is shown in Fig. 4.18 with an example constellation of parameter points. Here the coding ranges from a value of –2 for a 'strong adverse interaction' to a value of +2 for a 'strong useful interaction', with a value of zero for a 'neutral interaction'.

There is a significant difference between this method of coding and the one used earlier. Naturally, the subjective coding illustrated in Fig. 4.18 serves a different purpose in that it expresses the benefit to engineering, as opposed to objective coding according to, for example, the energy involved in the mechanism per unit time. Later we will also be discussing energy coding of the matrix.

In fact, in the cause *vs.* effect plot shown in Fig. 4.18, there is clear engineering significance in the position of the points in the diagram and, as a general principle, we would wish to move the 'rogue' adverse points into more positive areas of the diagram and, in the process, move the average value further up the $C=E$ line. The overall strategy would be to move the mean up the $C=E$ line; individual tactical engineering objectives would be dictated by the location of individual parameter points. The key point about Fig. 4.18 is that the engineering objectives can be incorporated via the matrix coding method, with the cause *vs.* effect plot indicating the significance of different engineering procedures.

The coding of the two 12x12 matrices presented in Chapter 3 could be according to a direct mechanical influence of one parameter on another, or according to how significant the mechanism is in terms of our objective: these are entirely different coding philosophies. In Fig. 4.19, the temporary excavation made for constructing

Fig. 4.19 Temporary excavation for hotel construction in Istanbul, Turkey, showing that the design life of the rock engineering (via the objective) will be critical.

the Hotel Dipoli in Istanbul is shown. A retaining wall with anchors is being built at the boundary of the excavation to inhibit the possibility of any slope instability and consequential damage to the buildings in the background. This excavation was made in greywacke rock and the question was asked whether the retaining wall with anchors was necessary, or whether indeed the rock slope in this excavation would have been stable without these protective measures. There is a variety of parameters and mechanisms involved in the mechanics of this situation, but the over-riding criterion is the objective — i.e. the stability of the slopes for a two year period while the hotel complex is being built and until the space between the basement floors of the hotel and the excavation walls can be backfilled. This key criterion of the engineering objective would indicate that project-based matrix coding must form the essence of further development of matrix coding methods.

Another example is shown in Fig. 4.20. This illustrates the proposed development of an open-pit mine near a river in Wisconsin, USA (this is a proposal and not a photograph of an actual mine). Naturally, the main criterion of this operation will

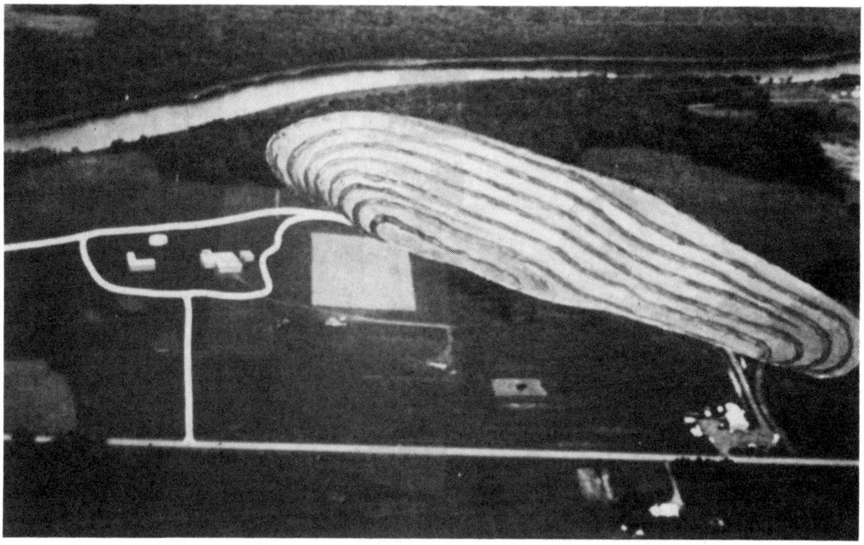

Fig. 4.20 Proposal for an open-pit mine, illustrating that there can be a variety of engineering objectives associated with the same project, Wisconsin, USA.

be financial: to maximize profits. There will be the direct financial analysis, but there will also be the rock engineering considerations. Is the mine likely to collapse in the manner already illustrated in Fig. 1.7, or is there any possibility that water from the river could find its way catastrophically into the mine.

There are various 'catastrophe scenarios' in all such engineering operations and these increase in the severity of their significance as the scale of the project increases. This is not only because more rock volume and more money is involved, but also because larger projects are more unstable in a systems sense. We could list out the parameters and the mechanisms and conduct 'catastrophe scenario' analyses to evaluate the unacceptable combinations of parameters. This latter example illustrates perfectly that the coding would have to be both in terms of the fundamental mechanics and the project objective.

4.5 THEOREMS 1 & 2

Finally, in this Chapter, two theorems are presented which have been implied in the earlier discussion. The first is Theorem 1: the mean of the cause co-ordinates is equal to the mean of the effect co-ordinates, and hence the mean of all the parameter points always lies on the $C=E$ line. Theorem 1 is illustrated in Fig. 4.21.

Each cause co-ordinate is the sum of a row. All the cause co-ordinates added together are all the boxes in the matrix added together. Similarly, each effect co-ordinate is the sum of a column. All the effect co-ordinates added together are also all the boxes in the matrix added together. Hence, the sum of all the causes is equal to the sum of all the effects and, when divided by the matrix dimension, the resultant mean cause equals the resultant mean effect.

Thus, whatever the constellation of parameter points in the cause *vs.* effect diagram, the centre of gravity of the points will always lie on the $C=E$ line. For this reason, Theorem 1 assists in the interpretation of the cause *vs*. effect diagram. For example, if one attempts to alter the position of any parameter point in the diagram, other points are bound to be affected in order that the mean position is maintained on the $C=E$ line.

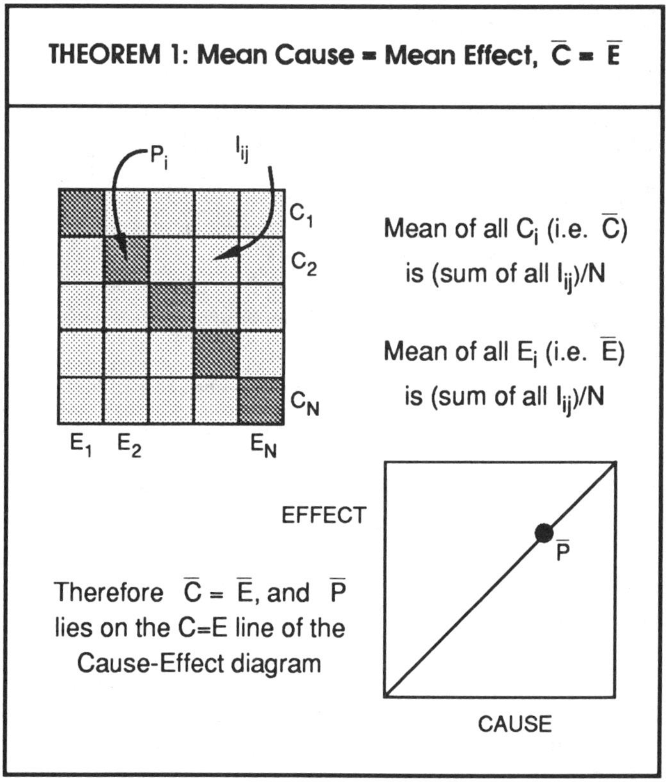

Fig. 4.21 The mean position of all the parameter points (i.e. their centre of gravity) lies on the $C=E$ line.

Theorem 2 is more subtle but is geometrically evident as illustrated in Fig. 4.22. The maximum parameter dominance or subordinacy potential occurs when the parameter is 50% interactive. As is clear in Fig. 4.22, at the origin there is no possibility of parameter deviation from the $C=E$ line because no mechanisms are operating. At the other end of the $C=E$ line, when all mechanisms operating in the off-diagonal boxes have maximal coding values, the system is 'saturated' and, again, there is no possibility of a parameter deviating from the $C=E$ line: they would all plot at the top right hand corner of the diagram in Fig. 4.22. It is clear from the diagram that the maximum deviation, represented by the perpendicular line from the $C=E$ line to the parameter point, reaches its maximum potential value when it originates from the centre of the diagram, i.e. when the parameter's interactive intensity is 50% of the maximum possible interaction.

Those readers with a background in stress analysis might be interested in the analogy between Theorem 2 and the fact that the maximum shear stress occurs at 45° to the directions of the two principal stresses. This is illustrated in Fig. 4.23. The shear stress represents an interaction between the normal stresses: given a specific stress state expressed by the principal stresses, only certain values of the shear stress are possible (note also that the shear stress plots in the off-diagonal

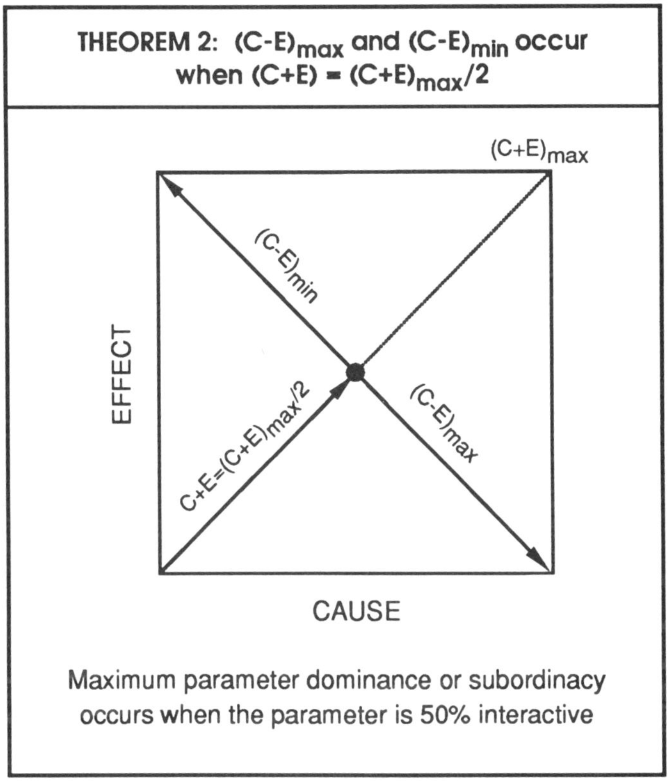

Fig. 4.22 Parameter dominance or subordinacy is represented by the point's distance from the 45° line, reaching a maximum from the centre of the $C=E$ line.

positions of the stress matrix). When the stress state 'deviates' from purely normal stresses, i.e. from the principal stresses which have, by definition, no associated shear stresses, the Mohr's circle diagram in Fig. 4.23 shows that the shear stress reaches a maximum halfway between the minimum and maximum principal stresses.

With the background of these two theorems, it should be remembered that there is enormous significance in the two characteristics of parameter interaction intensity and dominance, and how they occur together. We know from Theorem 1 that the constellation of parameter points must have its centre of gravity on the $C=E$ line. Within this constraint, there are many constellations that could occur, the two main ones being mainly along the $C=E$ line or mainly along a line perpendicular to it. If the parameter points are scattered along the $C=E$ line but fairly close to it, then

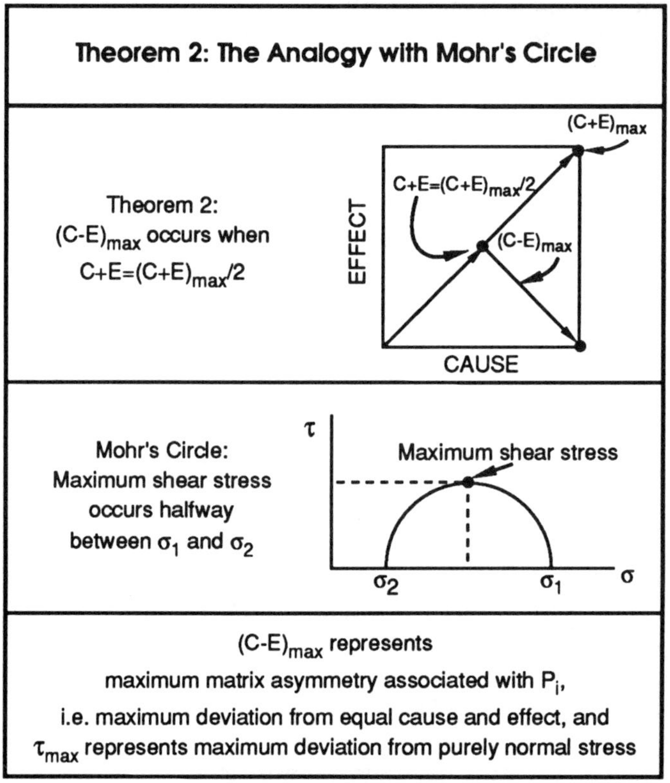

Fig. 4.23 One interesting analogue between Theorem 2 and stress systems is that the shear stress maximizes halfway between σ_1 and σ_2.

they can be ranked according to their parameter interaction intensity: in other words, they can be listed in order of interactive importance. If, on the other hand, the parameter points are scattered about a line perpendicular to the $C=E$ line, they will have similar interaction intensities but widely differing dominance values.

These possibilities have strong ramifications for rock mass classification schemes. In the former case, it might be possible to use, say, five or six parameters in such a scheme; in the latter case, all the parameters would have to be used. We will return to this theme in Section 9.2.

5

Examples of Matrix Coding and Cause–Effect Plots

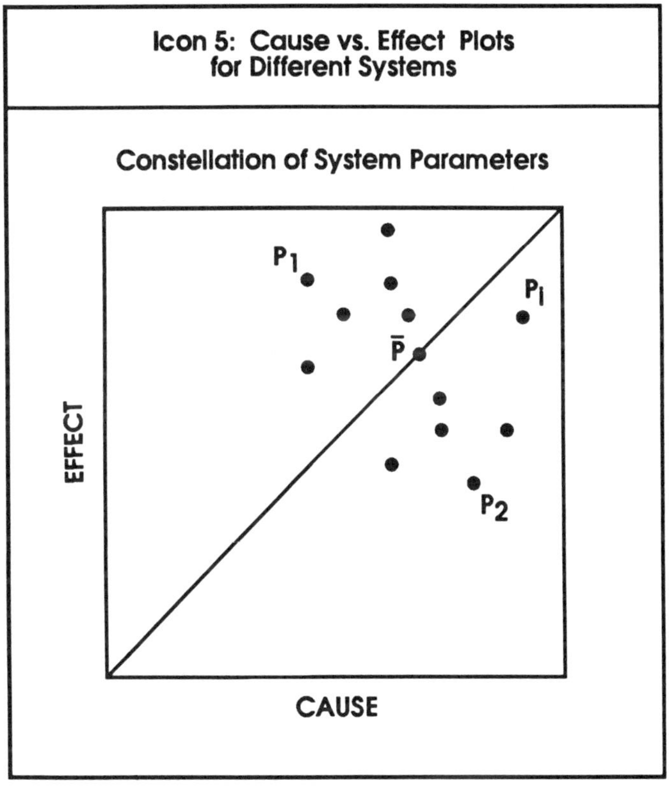

Each rock engineering system is characterized by a parameter constellation in cause–effect space.

The method of matrix coding and associated consequential cause *vs.* effect plots for interaction matrices are illustrated here by using the generic slopes and underground excavations matrices presented in Chapter 3. The second matrix coding method illustrated in Figure 4.3 (the ESQ method) will be used, i.e. an integer in the range 0 to 4 is allocated to each off-diagonal term.

5.1 ROCK SLOPES

In Chapter 3, the Atlas of Rock Engineering Mechanisms was presented. This contained interaction matrices representing generic slopes and underground excavations systems. Both these were 12x12 matrices, having different leading diagonal terms to reflect some of the different factors involved. In Chapter 4, the

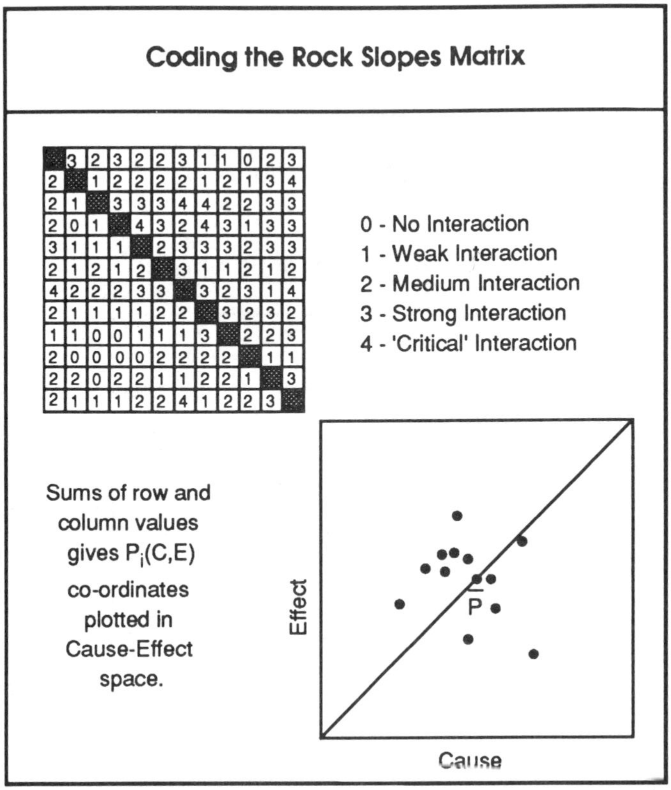

Fig. 5.1 Coding values for the generic slopes interaction matrix presented in Chapter 3 and the associated cause *vs.* effect plot.

method of numerically coding any matrix was discussed, and it was noted that there is a spectrum of methods varying from the simple binary on-off coding to a more explicit coding via numerical analysis. At this stage, the integer values of 0 to 4 seem most appropriate for an overall assessment of the scheme: this gives a sensitivity value from 0 to 44 for the summation of each matrix row and column, with a maximum value of 44 on each axis of the cause *vs.* effect diagram. When coded, these matrices each produce a constellation of parameter points as illustrated in Icon 5 at the beginning of the Chapter.

In Fig. 5.1, the coding of the rock slopes matrix presented in Fig. 3.1 is given, utilizing the second of the coding methods illustrated in Fig. 4.3. At the top left of Fig. 5.1, the individual coding values for the boxes are shown. For example, the box in the second column of the first row, i.e. Box 1,2 is the influence of the overall environment on the intact rock quality. In the Atlas (see Fig. 3.1a), this has the illustrated text "Type and state of rock quality depend on geological history". The box has been generically coded with a value of 3, i.e. there is a strong interaction because the overall environment will generally have a strong influence on the intact rock quality.

As can be seen in Fig. 5.1, all the individual boxes in the slopes matrix have been coded in similar fashion. In some boxes there is essentially no interaction, and a value of 0 is assigned to the box, e.g. in Box 1,10 which is the influence of overall environment on proximate engineering: "Proximity to tectonic blocks can cause failure". In other cases, the box has been coded with the highest value of 4 to represent a 'critical' interaction. For example, Box 3,8, which is the influence of discontinuity geometry on slope orientation and location: "Discontinuity orientations can dictate slope orientations".

The reader might like to check some of the other interactions in the Atlas in Section 3.1 to obtain a feeling for the coding procedure and the values given in Fig. 5.1. Naturally, there is latitude in the way that the coding numbers are applied. Moreover, the examples shown in the off-diagonal boxes in the Atlas are only illustrative of mechanisms that could occur in that box and are certainly not comprehensive. Also, the Atlas in Chapter 3 is 'generic': it is much easier to code the matrix when the project is known (together with the objective) and the mechanisms are more readily identifiable. The author has found that, in the vast majority of specific cases, values can be assigned to within plus or minus 1. Thus, the summed values of the 11 interaction terms in each row and column will not vary too much if some of the individual coding values were to be altered. In the future, such variability in the assignment of coding values could be incorporated via fuzzy arithmetic, for example, but the purpose now is to explain the method.

The individual rows and columns for the matrix are summed, as shown in Section 4.1 and Fig. 4.4, and the individual parameter cause–effect co-ordinates thus generated. These co-ordinates are used to plot each parameter point in the cause *vs.* effect diagram as shown in Fig. 5.1 and 5.2, e.g. P_1 has co-ordinates (22,24).

Note that the 'cause' refers to the influence of the parameter on the system and the 'effect' refers to the influence of the system on the parameter. It was mentioned earlier that the words 'cause' and 'effect' are reversible and depend on the rotational convention assumed in the matrix. Because a clockwise rotation has been adopted, the 'cause' is found from the sum of the *row* codings and the 'effect' found from the sum of the *column* codings. These would be reversed had an anti-clockwise convention been adopted for the matrix.

The cause *vs.* effect diagram for the generic rock slopes matrix presented as part of the Atlas in Chapter 3 is shown in the lower part of Fig. 5.1, initially, without any numerical indication of the co-ordinates. We find that the mean parameter value lies very close to the centre of the diagram, indicating that the rock slopes matrix represents a system that is about 50% interactive. There is a reasonable distinction between the highest and lowest level of parameter interaction intensity, remembering that this is indicated by the distance from the origin along the diagonal of the diagram to the parameter point. Also, there is a reasonable difference between the extreme values of parameter dominance and subordinacy, remembering also that these are represented by the perpendicular distance from the diagonal line in the plot to the parameter point. (The reader is referred to the Icon at the beginning of Chapter 4 and the expanded Icon in Fig. 4.15 for these concepts.) To interpret the complete constellation of parameter points, we will require more experience in the types of parameter configurations that can occur in such diagrams.

The cause *vs.* effect diagram is amplified in Fig. 5.2 with the axes numbered and the individual points identified. The parameter with the highest interaction intensity

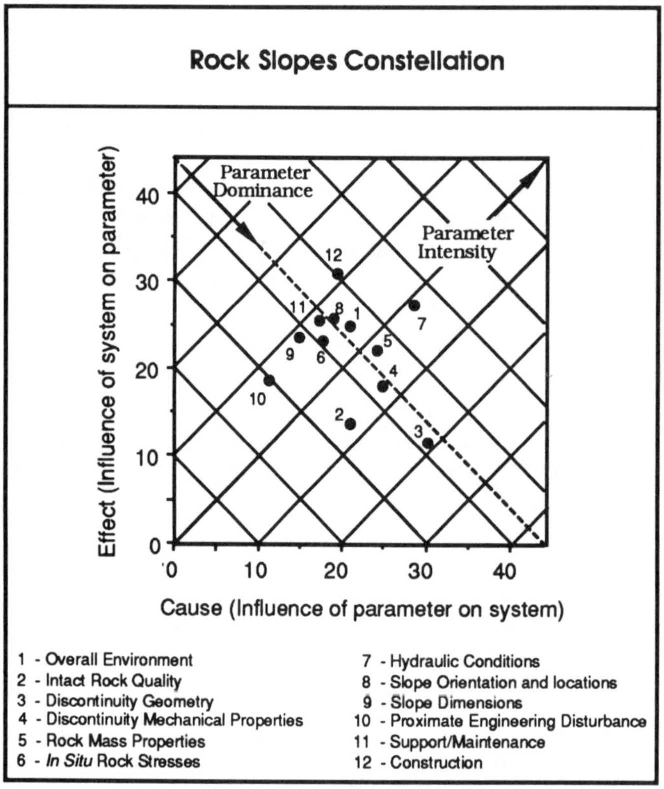

Fig. 5.2 Cause *vs.* effect plot for the generic 12x12 slopes matrix presented in Fig. 3.1, for the coding values given in Fig. 5.1.

is number 7, Hydraulic Conditions; the parameter with the lowest interaction intensity is number 10, Proximate Engineering Disturbance; the most dominant parameter is number 3, Discontinuity Geometry; and the most subordinate parameter is number 12, Construction. All these agree with our experience. We expect the hydraulic

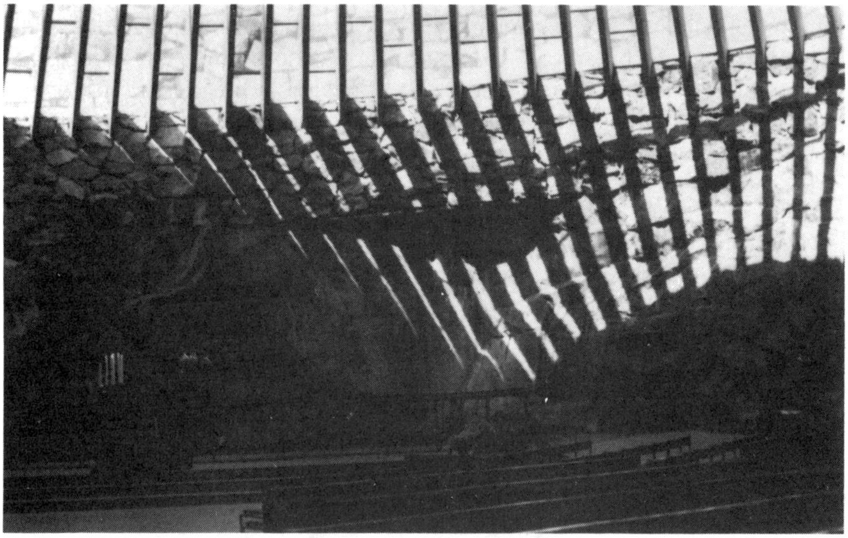

Fig. 5.3 Slope stability around the walls of a church excavated directly in granite in Helsinki, Finland.

conditions to have a severe affect. We would expect proximate engineering disturbance to be mainly, although not completely, decoupled from our system. Discontinuity geometry will always be important for slope stability and in this case is, in fact, the most dominant parameter, i.e. the parameter which has the greatest effect on a rock engineering project involving slopes. Finally, the construction itself, is the most subordinate parameter. This is also to be expected since we

Fig. 5.4 One of the many road cuttings along the Autopista highway A-7 south of Valencia, Spain.

design the construction on the basis of the rock, site and project conditions: in other words, all the other parameters are specifically accommodated in the design procedures. In the slopes interaction matrix, Construction is the twelfth box in the matrix. Note that the right hand column of numbers in Fig. 5.1 has high values because this column represents how all the individual parameters will affect the

Fig. 5.5 Blasting damage in a rock slope causing changed rock conditions and possible instability.

construction, i.e. these are the main design considerations. The constellation of parameter points in Figure 5.2 represents a relatively dense cluster and we can think of the parameters which we have just highlighted (hydraulic conditions, proximate engineering disturbance, discontinuity geometry and construction) as being the limits of the constellation.

In Figs. 5.3 to 5.6, there are four photographs of different rock slopes illustrating different aspects of slope instability. These are included to show how the generic atlas for slopes can be considered more specifically in terms of the differing objectives for specific engineering schemes. The matrix can then be coded accordingly. This recoding would tailor the matrix to the particular rock, site and project and assist in our overall consideration of the mechanisms that might be involved. Then, directly via the matrix or indirectly via the cause *vs.* effect plot, decisions can be

Fig. 5.6 Stable pre-split slope by the A82 road on the west of Loch Lomond in Scotland. (Note that the slope direction is continuously changing as the road bends.)

made regarding the optimal strategy for the given circumstances.

The first of the these illustrations, Fig. 5.3, shows a church in Helsinki which was created simply by excavating the surface rock and placing a wood and glass roof over the excavation — to dramatic effect. However, we would naturally wish to ensure the stability of the side walls of this church and would be concentrating very explicitly on the individual block failures which are controlled by the rock

Fig. 5.7 The Panama Canal, with the Pacific to the left and the Atlantic to the right. The black hill (Gold Hill), in the right middle centre, is the Continental Divide. Note the 1986 Cucaracha landslide in the foreground.

mass structure.

The Autopista highway A-7 near Valencia in Spain is shown in Fig. 5.4 at a location where it passes through one of the very many cuttings along its route. These cut slopes look extremely ‘clean’ and the emphasis here would be not only on the possibility of plane, wedge, or toppling failure but also on the long term stability of the slopes cut in this degradable limestone-type material. Thus, there is a similar safety criterion, but the rock type would have a stronger dominance because of the presence of karst features and highly variable weatherability characteristics.

The damage that can be caused by blasting is shown in Fig. 5.5 where an extremely unstable rock slope is shown. The dilation introduced by blasting is clearly visible. Here the influence of construction on the rock structure is of extreme importance in stability assessment. Moreover, the instability potential of the slope is now clearly more dependent on the post-construction properties of the rock structure

Fig. 5.8 The 1986 Cucaracha landslide (left centre) on the bank of the Panama Canal. (Atlantic to the left, Pacific to the right).

than on the pre-construction properties, highlighting a point that will be made later: it is not possible to establish completely the rock properties prior to construction, because they will be affected by the construction process.

A pre-split rock slope on the west bank of Loch Lomond at the side of the A82 road in Scotland is shown in Fig. 5.6. Here the road curves through an azimuth of about 90°. The dip angle of the slope had to be varied, initially to avoid plane failure and later to avoid wedge failure, as each failure mode became a potential hazard at different slope orientations. It was also necessary to implement drainage measures and to install rockbolts at some locations. One can see that the system is becoming more interactive and more complex in this latter example.

Consider the slope failures along the Panama Canal shown in Fig. 5.7 and 5.8. These are perhaps at the extreme range of interactive complexity. The geomorphology of this area is extremely complex, varying from hard basalts through a whole spectrum of weathered materials to soil-like conditions. There is a very strong effect of rainfall, with meteoric water passing through the rock structure. In fact, the best empirical indicator of failure on the banks of the Panama Canal is the 60-day cumulative rainfall value. There are pre-existing slips. There is the fluctuating level of water in the canal. The rock is very inhomogeneous and anisotropic, so that measurements of displacement and water heads are difficult to decode. All this indicates the extremely interactive nature of the rock engineering circumstances in that region.

The suite of photographs in Figs. 5.3 to 5.8 illustrates a progression of 'interactive complexity': the spectrum from the simple, and readily identifiable possibility of individual block failure in the walls of the granite church in Helsinki through to the very complex nature of the interactive mechanisms along the Panama Canal. The nature of these different slopes would be reflected in the constellation of parameter points in the cause *vs.* effect diagram, once the slopes interaction matrix had been coded separately for each slope and the associated circumstances.

5.2 UNDERGROUND EXCAVATIONS IN ROCK

In this Section, we follow the same procedure but consider the second 12x12 interaction matrix in Chapter 3. The matching diagram to Fig. 5.1, but this time for underground excavations, is Fig. 5.9. In the top left of the Figure, the matrix coding values are given according to the ESQ method (see Fig. 4.3). We can again highlight some of the extreme values to indicate the coding method. For example, Box 1,9 is coded as 0. This is the influence of cavern dimensions on intact rock quality. There could be some minor effect in that larger caverns might cause a greater degradation of the intact rock quality but, within the resolution of the coding, we would assign this box a value of 0. At the other end of the coding range, we can note that Box 2,10 has been assigned a value of 4, i.e. this is a 'critical' interaction, being the influence of rock support on rock behaviour. The whole purpose of rock support is to control the rock behaviour, as illustrated in Box 2.10, and so the coding must be 4.

The associated cause *vs.* effect plot in the lower part of Fig. 5.9 shows some interesting differences when compared with the corresponding plot for slopes shown in Fig. 5.1. The mean interaction intensity is higher and the parameter dominance and subordinacy has become stronger. Remembering that the interaction intensity is indicated by the distance up the $C=E$ line, we note that there is a close equivalence for almost all the parameters in terms of their interaction in the rock engineering system. However, the parameter dominance or subordinacy (indicated by the length of the perpendicular from the $C=E$ line to the point) is very variable. This is particularly interesting and significant, given that it could not have been predicted beforehand — either intuitively or by a cursory examination of the coding values for the two 12x12 matrices.

The cause *vs.* effect plot for underground excavations is clarified in Fig. 5.10 with the individual parameters identifiable. In this plot, we find that the most interactive parameter is number 3, the Depth of the Excavation. The least interactive parameter is number 6, the Discontinuity Geometry. The most dominant parameter is number 7, the Rock Mass Structure. The most subordinate parameter is number 10, the Rock Behaviour, which we would expect because this is conditioned by all the other parameters. It is emphasized that these are general conclusions about the nature of underground excavations as determined from the generic matrix. If faced

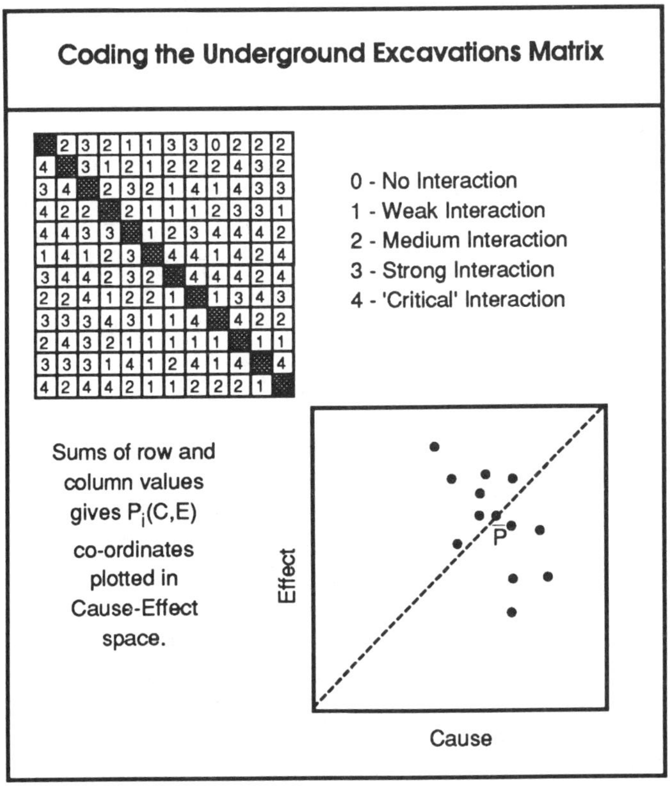

Fig. 5.9 Coding values for the generic underground excavations interaction matrix presented in Chapter 3 and the associated cause *vs.* effect plot.

with a specific rock type, a specific site and a specific project objective, the generic matrix could be coded accordingly. Naturally this would cause alterations to the positions of the parameters.

In Figs. 5.11 to 5.13, three aspects of underground excavations are illustrated. In Fig. 5.11, there is a photograph of an old unlined shaft at the Parys Mountain Mine on Anglesey in the UK. In this particular rock, there is essentially only one set of discontinuities. In terms of rock blocks, two adjacent discontinuities could form two surfaces of a block and the excavation surface could form a third. However, the minimum number of faces required to form a block is four, i.e. a tetrahedron. Since no block can be formed, no block failure can occur. Indeed, the shaft has been stable for many years. One can anticipate how the atlas matrix for underground excavations could be coded specifically for these conditions. The results would be very different to the overall approach shown in Fig. 5.10.

The longwall method of coal mining (see Fig. 5.12) is used in many countries. Inherent in this mining method is the use of controllable supports in an integrated face. Again, with this very specific method of what is effectively creating a moving opening in a coal seam, and with the emphasis very much on strata control, the matrix would be coded differently. It should be noted that, even with such a well-defined procedure as longwall mining, the interaction matrix would alert the designer to many mechanisms which previously might not have been taken into account.

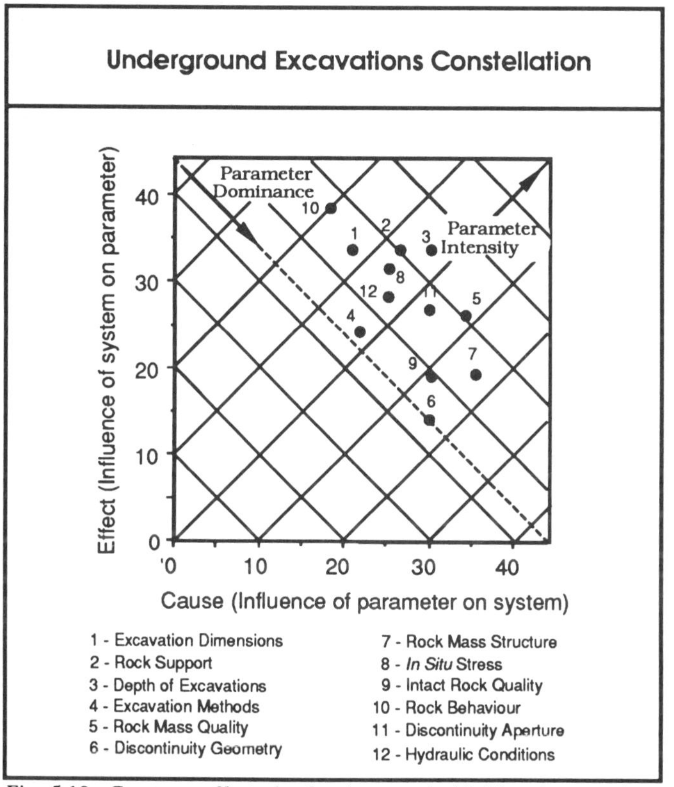

Fig. 5.10 Cause *vs.* effect plot for the generic 12X12 underground excavations matrix presented in Fig. 3.2, for the coding values given in Fig. 5.9.

Indeed, British Coal have recently implemented the idea that by aligning the advancing direction of the longwall face and the longwall face itself with two of the principal stresses (assumed horizontal) the rock damage effects are reduced, all other factors being equal. Similarly, British Coal is now progressing with enhanced use of rock bolts, which would also have been an output from the interaction matrix coded for coal mining.

Another aspect of Fig. 3.1, which is highlighted in Fig. 5.13, is Box 4,8 in the underground excavation portion of the Atlas. This is the influence of excavation

Fig. 5.11 Old shaft at the Parys Mountain copper mine in the north east of the island of Anglesey, Wales.

methods on *in situ* stress. The computer simulation in Fig. 5.13 represents stoping in highly stressed rock in Canada and associated seismic events. Excavation will always significantly affect the stress because all unsupported excavation surfaces are principal stress planes, having no shear stresses acting on them. In extreme cases, the excavation can cause such a dramatic alteration in the stress that the whole excavation is prejudiced. This will occur in regions which have a naturally occurring high *in situ* stress and/or mines with a high extraction ratio.

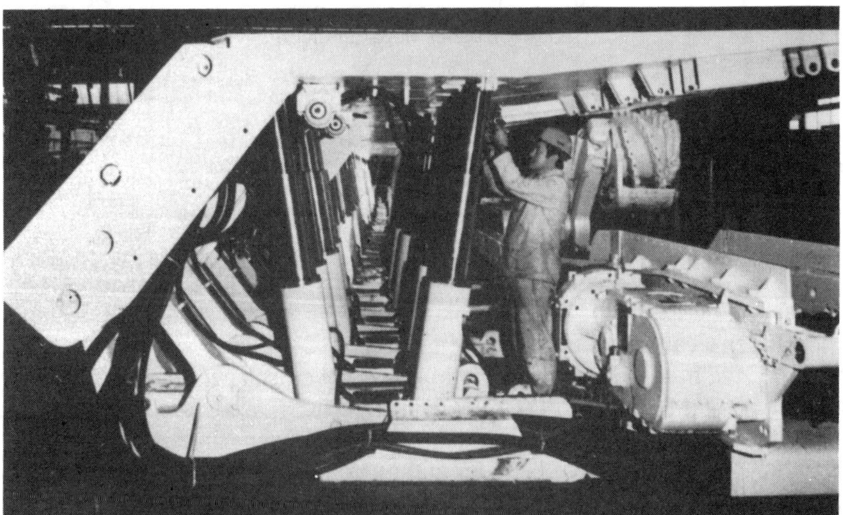

Fig. 5.12 Hydraulic supports used in mechanized longwall coal mining. (Systems approaches are now being adopted to design 'integrated face' mining methods.)

A rockburst occurred on Level 4 of the El Teniente Mine in Chile during the undercutting and development for a 300 m x 300 m block caving operation. Failure on many of the discontinuities occurred in a domino fashion, completely destroying the functional integrity of that region of the mine. This rockburst and others that occur in the mine are a function of the interactions between the complex rock mass structure, the extremely high stress field (caused by the subduction zone off the

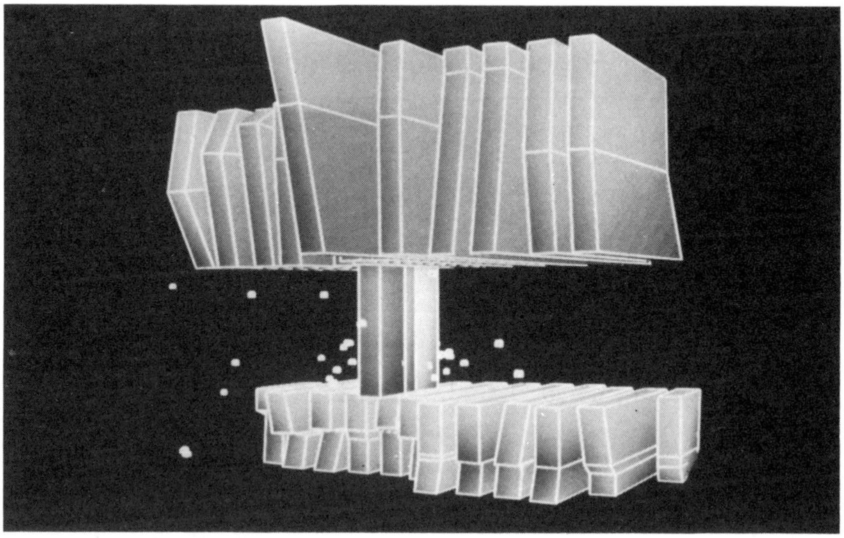

Fig. 5.13 Computer presentation of seismic events triggered by mining stopes in highly stressed rock in Ontario, Canada.

coast) and the dramatically large mining operation. In fact, El Teniente is the largest underground mine in the world and suffers from the instabilities of scale.

Basically, the larger the structure, the more unstable it will be. This is because the stored strain energy available to drive the failure mechanisms is proportional to the cube of the rock dimensions; whereas the work being expended on failure, shearing the discontinuities, is proportional to the square of the excavation dimension. The circumstances at El Teniente are interactively complex and hence, when faced with the final end product of failure, it is difficult to decode precisely what has happened. From the mining point of view, instability has occurred. From the systems point of view, as we will see later, *the mine has become more stable.* This illustrates the way in which 'systems thinking' can assist significantly in rock engineering projects, the subject of Chapter 10.

If it is known that a certain mechanism is operating, e.g. that the removal of the rock in the mine stopes is causing failure, either in the intact rock or along the pre-existing discontinuities, then specific models can be generated to establish the consequences of the mechanism. This is illustrated in Fig. 5.13 which is a graphics output of the seismic events that occurred during stoping operations at an INCO mine in Ontario, Canada. By matching the observed events and the product of numerical modelling, we can obtain sufficient information to generate the 'transfer function' for this part of the matrix. If this is possible, then we would need to concentrate on other regions of the matrix to see whether compatible information can be obtained for them.

There is an alternative method for presenting the information in Fig. 5.10, which is analogous to the hydrostatic and deviatoric axes of stress analysis. This is via the co-ordinates (*C+E*, *C-E*), which are the sum and the difference between the

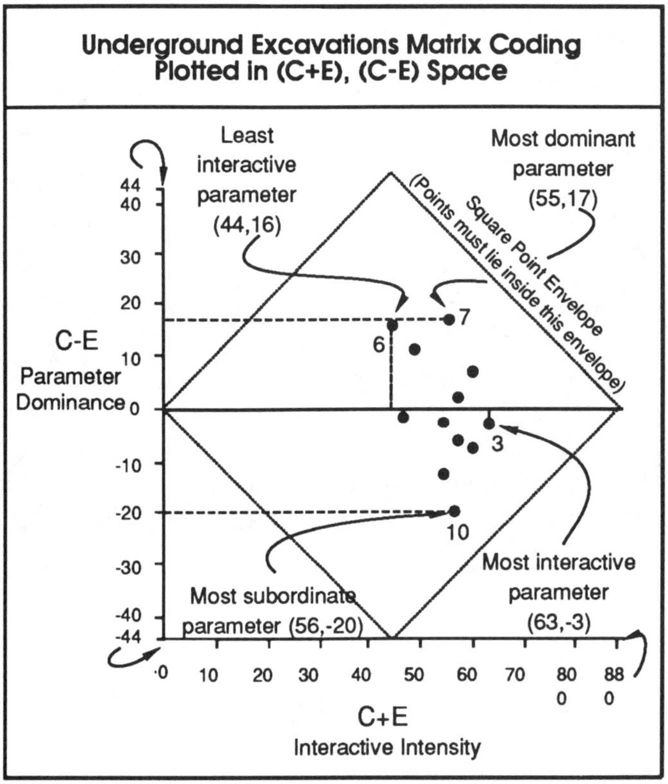

Fig. 5.14 Parameter points plotted in C+E, C-E space so that parameter interaction intensity and dominance can be seen directly.

totals of the row and column values passing through a leading diagonal parameter. These co-ordinates are presented for the 12 underground excavations parameters in Fig. 5.14. This Figure contains the same information as that in Fig. 5.10, but there is a more direct identification of the most interactive parameter (having the largest *C+E* value) and most dominant parameter (having the largest *C-E* value).

As the theory develops and the methodology is utilized more extensively in practice, it can be established whether the diagram in 5.10 or 5.14 is the most useful. In Fig. 5.10 the parameter points with their direct row and column positions are easy to understand. On the other hand, the sum of the row and column values and the difference between the row and column values shown in Fig. 5.14 are more diffficult to link back conceptually to the original matrix — but do provide the required information more directly, as shown.

It was mentioned in the Introduction to the book that rock engineering projects are becoming more complex, that we do have more information at our disposal, and that environmental issues are imposing increasing constraints on engineering activity. None of these is a problem within the systems methodology: in fact, they are actually all advantages in the broader context. However, a coherent methodology is required to optimize the engineering in relation to the objective.

One of the most striking examples that can be presented is that in Fig. 5.15, which illustrates the size of a proposed super-conductive energy storage magnet, shown next to the football stadium in Madison, Wisconsin, USA for size comparison.

Fig. 5.15 Proposal for a large super-conductive magnet to store electricity directly. Such a magnet would have to be located underground. (Wisconsin football stadium shown adjacent for size comparison.)

Traditional electricity generating stations operate optimally when they are producing a constant output. However, the demand for electricity fluctuates significantly throughout the twenty-four hour daily cycle. The highest demands for electricity in the UK occur during the surges associated with the ends of popular television programmes. A buffer is required to allow for the differences between the constant supply of electricity and the fluctuating demand. In the UK there are pumped water storage schemes in which water is pumped from a lower reservoir to an upper reservoir during the night when demand is least. This water is used to create electricity during periods of high demand. Depending on the hydroelectric system, there can be some minutes between the full pumping operation and the full generating operation. The concept behind the super-conducting energy storage magnet is the direct storage of DC electricity in the magnet, with only a few milliseconds switching time between output and input.

An alternative method of electricity storage is a super-conductive magnet. As can be seen in Fig. 5.15, the proposed magnet is enormous, the size of a football stadium. Because of its size and the need to cool it to achieve super-conducting capability, it has to be located underground. It could be either installed as one large magnet or as a series of magnets in tunnels. These tunnels would be circular in plan with magnetic flux passing through the rock to form one large composite magnet. Consider the interactions here. The magnet expands and contracts during charge and discharge cycles, resulting in a stress fatigue loading on the surrounding host rock mass. This has an amplitude of about 2 MPa. The intact rock may not be affected, but the discontinuities could well be. Given that the magnet can be constructed in sections, what is the best combined magnet–rock–construction system? What would be the effect of water on such a structure during its operation? What are the implications of thermal gradients in the rock induced by the cooling appparatus? To design and successfully optimize such a complex structure, utilizing the technical knowledge that we have and ensuring minimal environmental disturbance, would be a prime application of the methodology presented here.

Sometimes we may have to consider factors well outside our engineering knowledge. For example, what would happen to a bird flying over such a super-conductive energy storage magnet? More in line with our engineering thinking, what are the disaster scenarios for such a magnet? What are the unacceptable combinations of parameters in the matrix? How do we design against these?

This leads naturally to consideration of multiple rock engineering mechanisms operating concurrently and consecutively, and also to the need to interpret all the matrix pathways associated with these mechanisms. The method of approaching these subjects is discussed in the next Chapter.

6

Multiple Rock Engineering Mechanisms

Icon 6: Pathways Through The Matrix

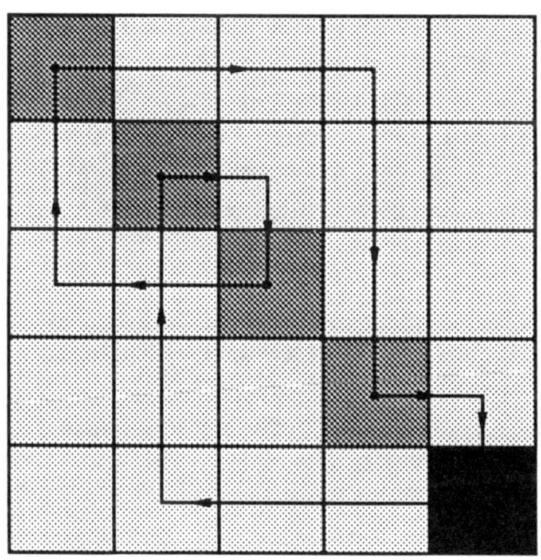

Suites of rock engineering mechanisms involving more than two parameters, operating concurrently or consecutively, can be represented by pathways through the interaction matrix.

Until now, all the interactions considered have been binary in nature, i.e. they have been the influence of Subject *A* on Subject *B* or the complementary influence of Subject *B* on Subject *A* and this concept of binary interactions has been used in Chapters 2 through to Chapter 5. Also, the cause *vs.* effect plot described in Section 4.2 and used for underground excavations and slopes in the last Chapter is based on linear addition of the coding values for the binary mechanisms.

A question that naturally arises concerns the representation of multiple rock engineering mechanisms of order 3 and above. One could study the influence of Subject *A* on Subject *B* and then include the influence of Subject *B* on subject *C*. This would be a three-fold or ternary mechanism and it would be possible to consider all possible extensions and variations on the theme, i.e. *x*-ary interactions where *x* is the number of parameters involved.

The idea of matrix resolution was discussed via Figs. 2.13 to 2.15 and it was noted with reference to Fig. 2.14 that it is possible to reduce compound mechanisms to binary mechanisms via the *AB(C)* concept as shown in Fig 2.14. However, at this stage to a much more flexible method is required for treating a number of mechanisms, operating concurrently or consecutively, and to study the general interpretation of a pathway through the matrix.

When the matrix is 'switched on', there will be binary and higher level mechanisms occurring simultaneously in the matrix. In other words, we establish the matrix to be in a given state and then consider the changes through an increment of time. For example, a composite elastic mechanism, which from the definition of elasticity has no time dependence, must occur instantaneously and several such mechanisms involving *x* parameters in total could operate concurrently. However, in practice most mechanisms will have some time dependency and therefore, after the matrix has been 'switched on', the interactions will become more complex and difficult to decode. So, as the matrix travels through time, it is also necessary to consider a number of mechanisms operating consecutively. In both cases, for concurrent and consecutive mechanisms, the composite mechanisms are represented by matrix pathways.

6.1 MECHANISMS OPERATING CONCURRENTLY

To illustrate the idea of a number of mechanisms operating concurrently, the idea of the 'Double Kick' is shown in Fig. 6.1. The four component steps in the development of the double kick are numbered 1 to 4 in the Figure. In the first step, the construction box is 'switched on', as represented by the black infilling of the last leading diagonal box. In the second step in Fig. 6.1, all parameters on the leading diagonal are potentially affected by the fact that the construction box has been switched on. This occurs through the interactions in the last *row* of the matrix. The first 'kick' occurs in the third step when the alteration in the leading diagonal parameter values affects the construction box itself via the interactions in the last *column* of the matrix. Thus, because of the original perturbation to the construction box, all the leading diagonal parameters are affected and these in turn affect the construction box. In other words, the system 'kicks back'.

Furthermore, once all the leading diagonal parameters have been affected by the initial perturbation in the construction box, they then affect one another via all the off-diagonal interactions in the matrix. Then, via the interactions in the last column of the matrix, the construction box is again affected: this is the second 'kick' and potentially a greater response. In Fig. 6.1, the boxes are shaded more darkly as they are activated. In reality, the matrix response to a construction perturbation

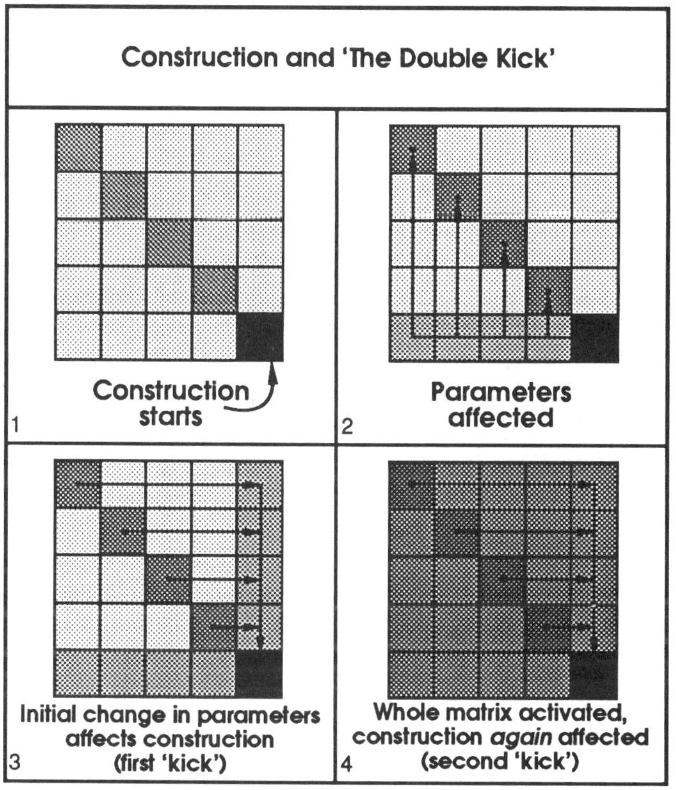

Fig. 6.1 The four stages in matrix evolution, from the initial construction perturbation to complete parameter interaction.

will not be as simple as a first 'kick' and a second 'kick'. This is because all the mechanisms are likely to have different time constants. For both the first 'kick' and the second 'kick', there will be a merging of all the mechanisms into a more or less continuous system response.

This raises the question as to how we might represent a complex mechanism such as the one intimated by Icon 6 at the beginning of this Chapter. What are all the pathways through the matrix and how do we represent any number of mechanisms operating concurrently? To illustrate the pathway concept, the ternary interactions (i.e. the interactions involving 3 leading diagonal parameters) are shown in Fig. 6.2, where the path dependency has been indicated. If there were the 3 parameters, A, B and C as the leading diagonal terms, there would be six parameter permutations: ABC, ACB, BAC, BCA, CAB and CBA. The permutations ABC and CBA are complementary pairs in the sense that the path dependency has simply been reversed. Note that in Fig. 6.2 the six ternary interactions are shown as three sets of complementary pairs. Faced with the binary interactions discussed in Chapters 2-5 and then asked to extrapolate the concept to multiple parameter interaction, we might have been tempted to extend the square matrix to a cube for ternary interactions.

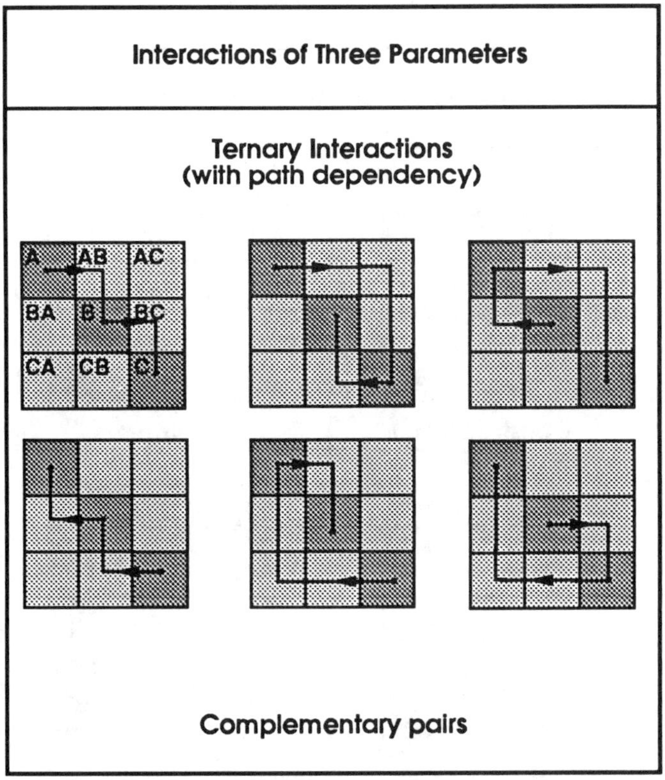

Fig. 6.2 Illustration of how ternary interactions (involving three parameters) can be represented by matrix pathways.

Mathematically-minded engineers might invoke hypercubes in x dimensions for mechanisms involving x parameters, i.e. for x-ary interactions. However, this is not necessary because the matrix pathway representation is simple and efficient.

One of the ideas that can be immediately studied through the pathway concept is the fact that, if there is effectively no mechanism or a zero-coded mechanism in a box on the pathway, then the multiple mechanism will simply come to a halt, as shown in Fig. 6.3. This means that the two parameters are essentially independent. A change in P_i will have no effect on P_j, i.e. $I_{ij} = 0$. Note that, because the matrix is asymmetric, it does not follow that P_j will not affect P_i, i.e. the complementary term I_{ji} may or may not be zero.

There are several possibilities to consider in the context of an 'empty box'. The first is that the off-diagonal box in the original matrix may be empty. In this case, there is an automatic termination for all potential pathways passing through that particular I_{ij}. An alternative is that the mechanism may be originally present but, due to a parameter threshold or mechanism threshold being reached during the mechanistic evolution of the matrix, the mechanism in question may cease to operate. The converse is, of course, also possible. The box may have an inoperative mechanism initially, but it starts to operate when the two associated leading diagonal parameters reach a 'mechanistically acceptable' combination for that box.

The density of off-diagonal terms that are inoperative will have a significant effect on the interactive intensity of the system and also the number of pathways

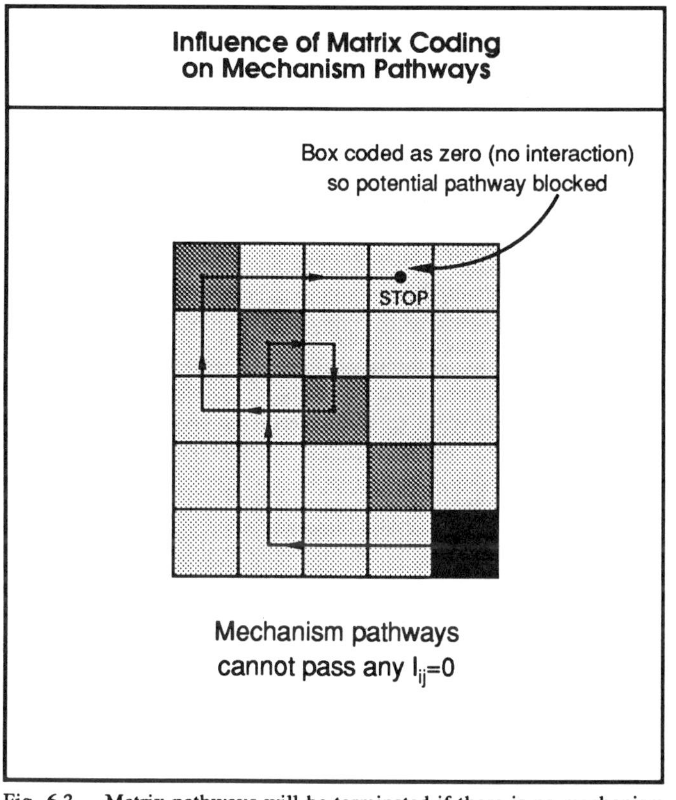

Fig. 6.3 Matrix pathways will be terminated if there is no mechanism linking the relevant leading diagonal parameters.

through the matrix — in a non-linear fashion. Very sparse matrices will have very few pathways passing through all the leading diagonal terms.

In Fig. 6.4, the matrix pathway idea is extended. One could start at any parameter and find a way through the matrix, possibly visiting every parameter, and hence establish a matrix mechanism pathway. In Fig. 6.4, a 2x2 and a 12x12 matrix are shown. There are only two possible pathways linking the two parameters in the 2x2 matrix but there are 12! pathways, i.e. 479,001,600 pathways, linking the 12 parameters. Utilizing appropriate computer algorithms, one could establish a variety of pathway characteristics which will be of interest. For example, what is the energy consumption of each pathway? Which pathways cause the parameter thresholds to be reached? Which composite mechanisms occur the fastest? Which parameter values are destabilizing and which parameter values are stabilizing the matrix? The answers to these questions will indicate the effects of engineering and the consequences of altering certain parameters, either directly or indirectly through the matrix pathways.

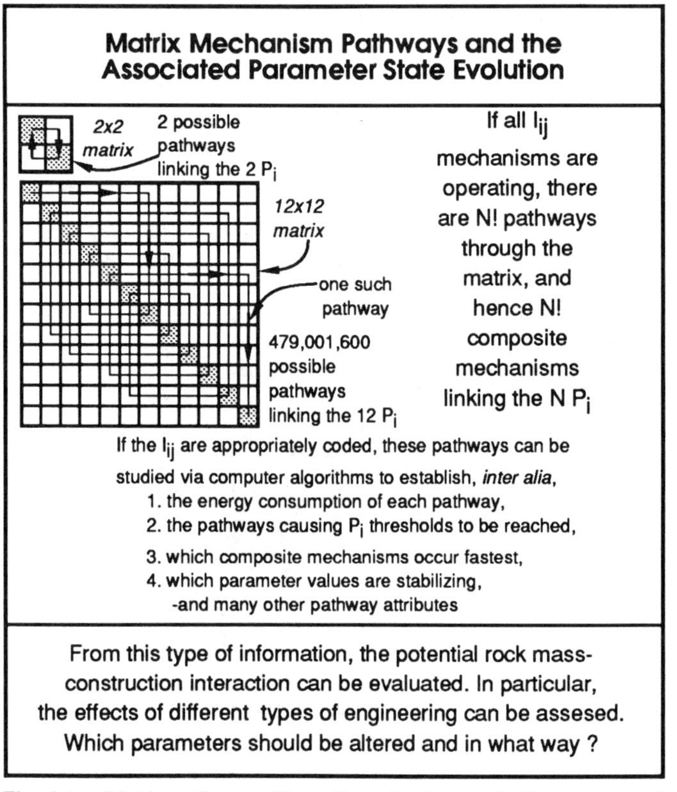

Fig. 6.4 Matrix pathways allow all mechanisms and all sequences of mechanisms to be represented.

The reader may have realized from the foregoing discussion that it is difficult to separate out mechanisms occurring concurrently and mechanisms occurring consecutively. Even though one pathway through the matrix will represent one concatenation of interactions, it will take time for these to occur. We can consider them as a single composite mechanism, with the path dependency indicated by the matrix pathway. However, by generalizing the pathway concept, it is quite possible for the matrix pathways to bifurcate and multiply. Thus the distinction between the concurrent and consecutive operation of mechanisms becomes blurred. As a general principle, therefore, it is probably best to simply consider the evolution of the matrix through time accounting for whatever multiple mechanisms are occurring. If no time is invoked, the mechanisms will occur concurrently; with time, they will overlap or be consecutive.

6.2 MECHANISMS OCCURRING CONSECUTIVELY

The author was once asked to evaluate the authenticity of an important fossil in the British Museum. This fossil has the name *Archaeopteryx lithographica*: the genus name '*Archaeopteryx*' means 'winged creature'; the species name '*lithographica*' refers to the fact that the Solenhofen limestone in which the fossil was found in Bavaria is used for making printing plates. *Archaeopteryx lithographica* is known to have been a small dinosaur. The specimen held by the Natural History Museum in London is shown in Fig. 6.5. Note the tail feather at the base of the specimen.

It has been suggested by critics that this fossil is a forgery and that the impression of the feather was created by imprinting a modern feather into a paste of very fine Solenhofen limestone particles. Although critics consider the fossil to be a forgery, the staff at the Natural History Museum are convinced that the fossil is genuine. The author was asked not whether the fossil was genuine but whether the *host rock*

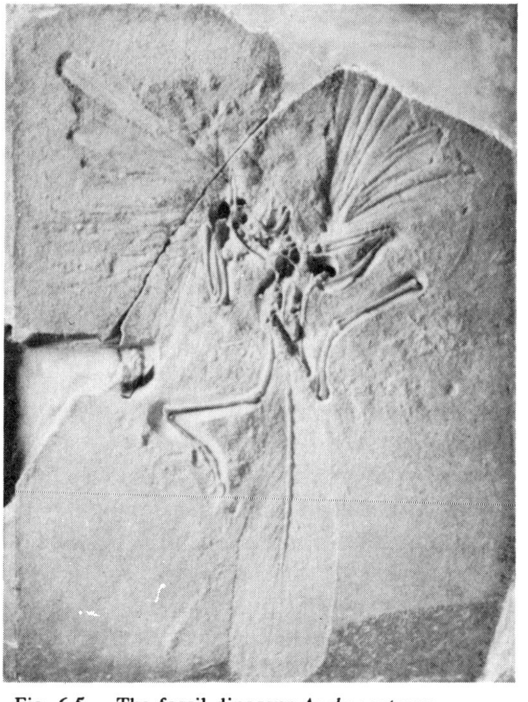

Fig. 6.5 The fossil dinosaur *Archaeopteryx lithographica*. Genuine or a forgery?

was genuine. A photograph of an enlarged section of the tail feather is included as Fig. 6.6. This photograph is of a region just a few millimetres across, the ridges being the separate filaments of the tail feather. Traversing these is a hairline crack in the Solenhofen limestone, travelling from the bottom left to the top right of the photograph. This hairline crack is filled with small calcite crystals. It is clear from the photograph that the feather impression must have occurred before the hairline crack. Calcite crystals were subsequently precipitated from circulating ground water.

Even with today's technology (let alone that available in Victorian times when the Natural History Museum purchased the fossil), it would be impossible to reproduce

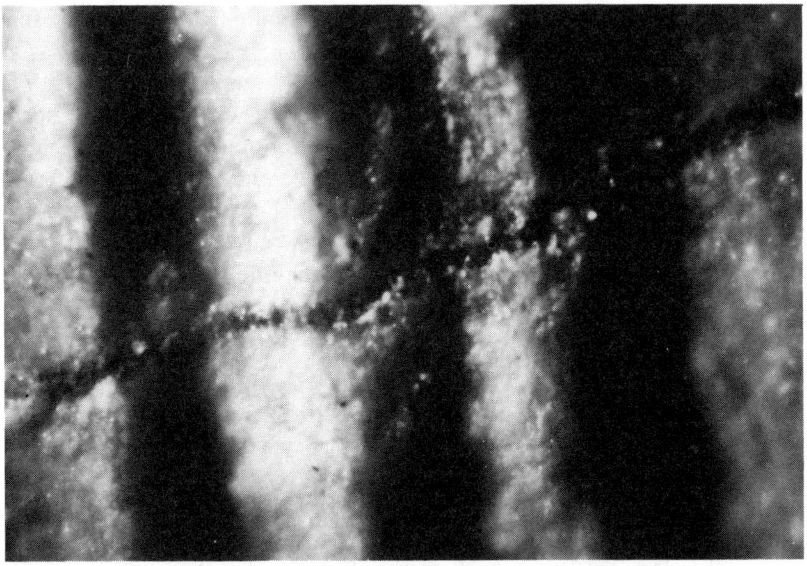

Fig. 6.6 Close up of tail feather of *Archaeopteryx lithographica*, showing infilled fracture.

this manifestation of natural geological processes and so the author is convinced that the fossil host rock is genuine — because of path dependency. This *Archaeopteryx* anecdote has been included to show how a knowledge of pathway mechanisms can assist in the interpretation of rock mechanics processes. In fact, using the matrix device and with extrapolation to all pathways, all rock engineering mechanisms can be studied.

In Fig. 6.3, the idea of a limit on the mechanisms was introduced, i.e. when a threshold is reached, a mechanism becomes inoperative. A pathway might be in operation for a certain time and then stop, either because the mechanism had attenuated or because a parameter threshold had been reached. This idea of a parameter threshold value is explored further with a 4x4 matrix in Fig. 6.7. For illustrative purposes in this Figure we consider all the mechanism pathways by which the first parameter on the leading diagonal can affect the fourth parameter on the leading diagonal. In the general context of multiple mechanisms, this interaction could occur directly, or via the second parameter, or via the third parameter, or via both the second and third parameter. These five interaction modes are illustrated in Fig. 6.7.

As a consequence of one of the indirect pathways involving one or two extra parameters, let us assume that a critical threshold is reached basically rendering the other parameter(s) inoperative. There are of course other possibilities but, for the sake of introducing the ideas, assume that all mechanisms in the *row* and *column*

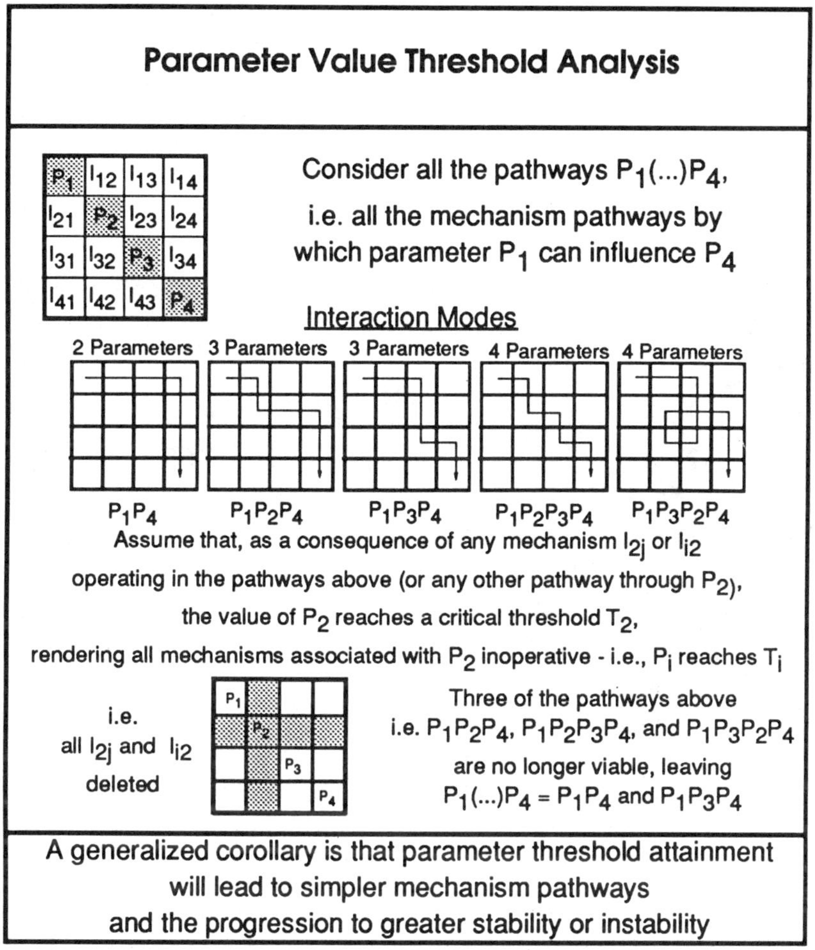

Fig. 6.7 As a result of the rock engineering mechanisms, parameter thresholds may be reached reducing the number of pathways.

through the <u>second</u> parameter are deleted. Under these circumstances, the number of interaction modes whereby the first parameter can influence the fourth parameter is reduced from five to two, as indicated in Fig. 6.7.

There are many interesting implications for engineering raised by this type of analysis. For example, it is likely that a generalized corollary of the concept in Fig. 6.7 is that parameter threshold attainment will lead to simpler mechanism pathways and the progression to greater stability or instability.

The photograph in Fig. 6.8 shows the concrete supporting pier of a car park on the island of Jersey in the Channel Islands. It can be seen that the foundation rock is highly fractured and has been rockbolted in order to improve the mechanical characteristics of the rock. In rock engineering, we do not necessarily have the 'go/ no-go' type of approach that is ubiquitous in other aspects of civil engineering, with the associated definite factors of safety. Whether this supporting pier is safe as a foundation element will depend on the rock movement and any consequences to the structure.

However, the associated rock–structure interaction will be representable by a diagram of the form shown in Fig. 6.7. The analysis must take into account what will happen if any threshold is reached, and whether the consequence of this is greater stability or instability. In many cases, it is not at all obvious. For the foundation pier illustrated in Fig. 6.8, some small movement on a discontinuity

Fig. 6.8 Foundation below a car park on the island of Jersey, UK, illustrating that pier movement may or may not lead to increased stability.

surface could cause unacceptable damage to the car park structure above. Conversely, such small movement may activate the resistance of the rockbolts and hence lead to greater stability. It is this type of analysis that needs to be conducted for the system and, indeed, all the sub-systems of a project.

An alternative example is illustrated in Fig. 6.9, which presents the side view of a Robbins 5 m diameter tunnel boring machine (TBM). The main horizontal ram visible in the picture (centre of the picture, halfway up the machine) is to provide the necessary thrust for the cutting head. The thrust plate facing the reader in the centre of the machine is the plate at the end of the reaction ram. When the TBM is in a tunnel, this plate is moved towards the reader to react against the tunnel wall at axis level.

The excavation process is a system and it has sub-systems. The system will operate successfully if the sub-systems are operating successfully. Some sub-systems are crucial to the operation of the total system, e.g. the main bearing of the TBM; other sub-systems are inconvenient if they fail but one can recover from them, e.g. a block falls from the roof. Finally, other sub-systems are not on the 'functional path' of the total system, e.g. small quantities of water entering the tunnel.

However, remembering Fig. 6.7, there is a variety of possibilities. If the cutting head encounters a rock stratum which it will not cut, the total excavation system breaks down. Similarly, but in relation to the other end of the rock strength spectrum, if the reaction ram penetrates soft rock and slips so that insufficient thrust is generated, the total system again breaks down. There is, therefore, a limited spectral window of rock strengths within which this tunnel boring machine system will operate without major failure occurring.

Considering such breakdown scenarios it may be that, rather than a single relatively clear threshold being attained, a more complex set of circumstances develops. For example, while the machine is passing through inhomogeneous ground, the rock at the cutting head has become harder but the rock near the reaction ram has become softer. Without either strength going outside the acceptable 'window' of rock strength, the combined rock strength values could cause a breakdown of the system. This latter case may not appear at first sight to be as serious as the two discussed earlier, but it could be. It would certainly require remedial measures — with the associated financial implications.

The reader may well be able to imagine other 'multiple mechanisms' involving fractured rock, difficulty with steering, jammed conveyors for example. There has always been the question with TBMs as to whether they should be made robust

Fig. 6.9 Side view of Robbins 5 m diameter full-face tunnel boring machine. If the rock is too hard, the TBM cannot cut; if the rock is too soft, the side plates will slip.

enough to be able to encounter successfully any ground conditions or whether they should be designed to the specific anticipated conditions. The former is more expensive but less risky; the latter is cheaper but more risky. This case example illustrates extremely well the whole question of mechanism pathways, the engineering implications, and the need to interpret the pathways via a coherent and structured methodology.

These ideas can be extended to 'parameter value analysis' as shown in Fig. 6.10. This term refers to the alteration in the values of the leading diagonal terms. The pathways associated with a particular concatenation of mechanisms will inevitably alter the parameter values. The new values may be acceptable or unacceptable, depending on the engineering objective. One only has to contrast the potential matrices and mechanisms associated with Figs. 6.8 and 6.9 to realise how much will depend on the engineering objective.

It is also important to note that the engineering objective of a sub-system and the total system will generally be different. In the radioactive waste disposal context, the overall objective is to ensure that unacceptable quantities of radionuclides do not return to the biosphere. The objective of the cavern emplacement sub-system

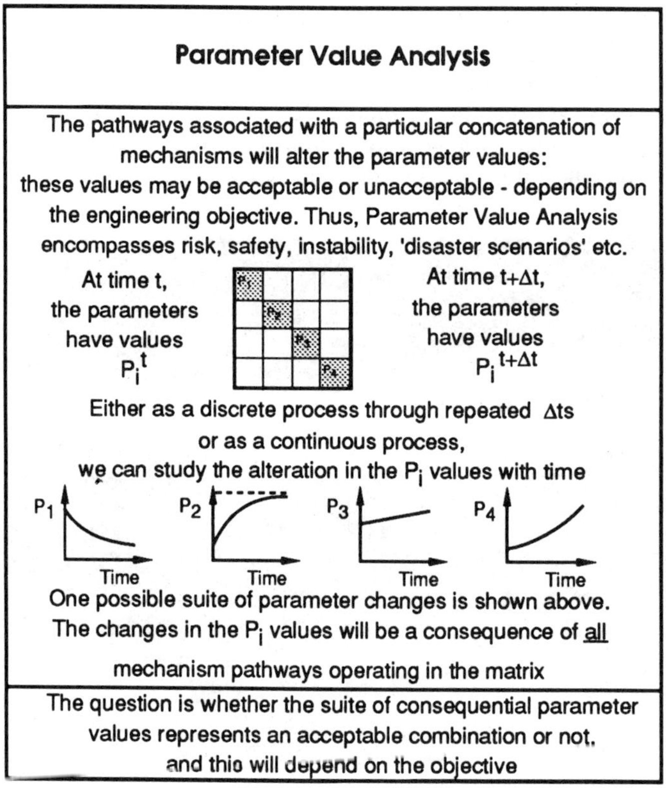

Fig. 6.10 As the mechanisms in the matrix evolve, the leading diagonal values change exhibiting a variety of behavioural modes.

might be that the cavern should remain stable for the design life of the repository. These are quite differing objectives, with quite different analyses.

In Fig. 6.10, different modes of parameter change as a consequence of various possible mechanism pathways operating in the matrix are illustrated. These include, for example, the following types of behaviour: negative exponential, increasing to an asymptotic value, linear increase and exponential increase. The fundamental question in parameter value analysis is whether the suite of consequential parameter values represents an acceptable combination or not, and this will depend on the objective. Parameter value analysis therefore potentially encompasses risk, safety, instability and 'disaster scenarios'. There is much work to be done in developing these analyses and it is hoped that the concepts presented here will form the foundation for the approach. It is also devastatingly apparent from the foregoing discussion that, unless this type of analysis is conducted, there is no hope of understanding the rock engineering — nor of conducting the engineering in an optimal fashion.

6.3 MATRIX PATHWAY INTERPRETATION

It has been seen in the earlier chapters that the compilation of an interaction matrix allows identification of all the binary interaction between the parameters. In the case of a 12x12 matrix, there are 144 boxes, of which 132 are off-diagonal boxes representing the binary interactions. The author contends that it is not possible to identify all these off-diagonal mechanisms and to consider their significance without the matrix device, or some equivalent procedure. The same applies to the rock engineering system. What are the multiple mechanisms and which are important for a given project?

The matrix pathway interpretation allows one to extend the systems approach to consider all the multiple mechanism potentialities and to assess their significance. Because rock engineering projects are becoming more complicated and because we have more and more computing capability at our disposal, the time is ripe to develop and utilize this systems approach. It enables the rock engineering design and construction techniques to be methodically derived, taking into account all possibilities.

Of course, in the general development of the associated methodology, it has to be possible to interpret the matrix pathways. In Fig. 6.11, we consider the 4x4 matrix which was first presented as Fig. 2.7, with the leading diagonal terms Rock Structure, Rock Stress, Water Flow and Construction. Assume that it has been possible to energetically code the off-diagonal components of the basic matrix for a particular project, i.e. an energy quantity has been assigned to each I_{ij} mechanism. Here we will now consider pre-split blasting for a highway project in the context of this energetically coded matrix.

The engineering perturbation is initiated in the construction box, i.e. C_{ij} on the leading diagonal. Considering the consequential pathways that can arise through the remaining leading diagonal terms, as they are affected one after the other, there are six pathways through the remaining three leading diagonal parameters of rock structure, rock stress and water flow. It is necessary to interpret the six pathways in engineering terms. (Note that the small table in the diagram indicates the units of energy flux in the mechanisms linking the permutated parameters. The energy coding used here is arbitrary for this illustrative example, the numbers being included to explain the principle.)

We can now identify which parameter permutation pathway has the highest energy and which pathway has the lowest energy level. These two pathways are indicated in Fig. 6.11 with 'candidate interpretations'. It can be seen that the high energy pathways would lead to the possibility of remedial action being required, whereas the low energy pathway interpretation might lead to little or no effect on construction. We would have to establish which of these six pathways was actually likely to happen, and which would lead to stability or instability.

As a general principle, the alteration of the construction box (and hence the possibility of alteration of all the parameter values) will lead to attenuation of the

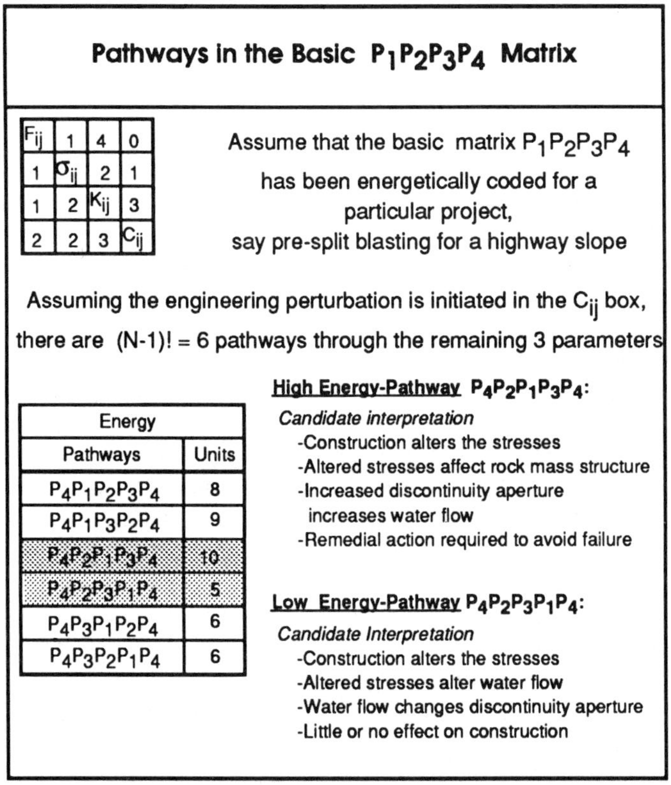

Fig. 6.11 Different pathways through the matrix have different rock engineering mechanism sequence interpretations.

mechanisms in the matrix and hence stability — otherwise every minor construction activity would lead to major disaster. However, within the systems context, we have a methodology, via the matrix pathway interpretation, of identifying risk, safety, instability, and 'disaster scenarios' for all projects.

It has to be remembered at this juncture that not all mechanisms are understood, or in some cases even partially understood. Referring back to the interaction matricess in Chapter 3, it is clear that we understand a great deal about how the rock structure affects the water flow, because the network of discontinuities dictates the secondary permeability of a rock mass. However, we know less about the complementary box: how does the water flow affect the rock structure? As water is flowing through discontinuities, how are they affected? Do they silt up after one hour, one day, one month or one year? Is the rock material eroded so that the secondary permeability is considerably enhanced with time? Does the channel flow that occurs in all fractures change with time?

In Fig. 6.11 candidate matrix pathway interpretations during pre-split blasting are presented. However, given the uncertainty expressed in the previous paragraph about binary mechanisms, there are infrastructural lack-of-knowledge problems concerning compound mechanisms. The process of pre-split blasting illustrates this point well. The purpose of pre-split blasting is to create a plane through the rock prior to excavation by simultaneously detonating low density explosive in closely-

Fig. 6.12 Simulated creation of a pre-split blasting plane through 'discontinuities' in Perspex.

spaced parallel boreholes. (An example of attempted pre-split blasting was shown in Fig. 2.21.) Is the pre-split plane formed by the very fast stress wave effect of the explosive, or by the relatively slower gas pressure effect?

In Fig. 6.12 there is an illustration of a laboratory experiment carried out in Perspex (polymethylmethacrylate) containing two simulated discontinuities. Three detonators were set off simultaneously to mimic site practice. In the lower part of the photograph, one of the detonator holes is clearly visible in the dark region, which is the shattered Perspex in close proximity to the detonation. The increased shattering caused by reflection off the free face at the discontinuity is clearly evident. Above this detonator hole is damage that has been contained between the two

Fig. 6.13 Pre-split plane formed by using slow acting expanding cement, Isle of Lewis, Scotland.

discontinuities because of reflection of the stress waves at the free surface. At the top of the picture, there is a mirror image fracturing effect similar to the hole at the bottom of the picture.

The purpose in showing this photograph is the demonstration that, even under these very adverse fracturing situations and with some additional fracturing introduced, a pre-split plane is actually obtained as a result of the stress wave effect. Thus, one might not expect the gas pressure effect to contribute significantly to the creation of the pre-split plane.

However, consider the photograph in Fig. 6.13 which shows a pre-split face on the Isle of Lewis in Scotland. The visible nature of the pre-split plane may appear to accord very well with the overall mechanics discussed with reference to Fig. 6.12 — but the pre-split plane illustrated in Fig. 6.13 was obtained by using a slow-acting expanding cement in the boreholes, and not an explosive. There was no stress wave effect; there was only a 'pressure' effect. So, how do the separate components of the stress wave effect and the gas pressure effect contribute to pre-split blasting? No one knows.

Thus, our development of this systems approach must begin with the knowledge that not all mechanisms are understood. We will generally know that one parameter will affect another in some way, but we will not always know the precise mechanisms. This is not, however, a barrier to the implementation of the systems approach. The intention is to provide a structured methodology using existing information to optimize engineering. Such uncertainty will become an inherent part of the systems approach,

Pathways Through the Matrix

There are a variety of existing problems and associated algorithms for establishing pathways - given a network and an objective; some of these* are 'translated' here in terms of the interaction matrix

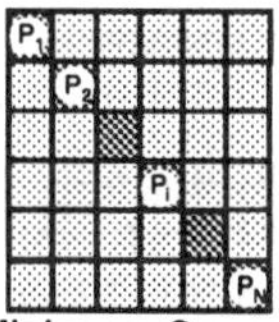

The Iij represent interactions between the Pi and Pj. In particular, they could be energy flows

1. Minimum Connector problem:
   What path through the I_{ij} links **all** the P_i parameters with minimum ΣI_{ij}?
2. Travelling Salesman problem:
   Starting at one P_i, say P_N, what route through **all** the other P_i and back to P_N is associated with minimum ΣI_{ij}?
3. Chinese Postman problem:
   Starting at one P_i, say P_N, what route through **all** the I_{ij} back to P_N minimizes the ΣI_{ij} (some I_{ij} may be zero)?

*The pathway problems are discussed in "Decision Mathematics" by The Spode Group, Ellis Horwood, 1986, 464pp.

Fig. 6.14 Relation between classical mathematical network problems and rock engineering matrix pathway mechanisms.

as it is with all systems, and we must develop our procedures to account for uncertainty via traditional statistical techniques, fuzzy arithmetic and algebra, or other techniques. Finally, it worth noting in this Chapter on multiple rock engineering mechanisms, that there may be some analogues between traditional mathematical network and pathway problems and rock engineering systems. Three traditional mathematical problems are shown in Fig 6.14.

The first is the 'minimum connector' problem. There are N locations in space. How can these be connected with telephone or electrical cables to minimise the connector length? Translating this into the present rock engineering context, the problem becomes which path through the off-diagonal terms links all the parameters with a minimal coding, representing some objective, say minimal energy.

The second is the 'travelling salesman' problem. If there are N towns, what route through all the towns minimizes the distance a travelling salesman will have to travel? There is no closed form solution to this problem: it has to be solved algorithmically. In our current context, this would represent a multiple mechanism in which one started at a specific parameter, and asked what route through all the other parameters back to the starting position would be associated with a minimum summation of the coding values in the off-diagonal boxes passed through.

The third problem is the 'Chinese postman' problem. This is similar to the travelling salesman problem except that the poor postman has to visit, not only every town, but every street and every house. Which is the route that minimizes this distance, achieved by minimizing any duplication of travelling? In the rock engineering context, this represents initiation of a pathway at one parameter and then travelling back through all the off-diagonal terms to the original parameter, and minimising the summation of the coding values.

These three traditional mathematical problems all have interesting analogues in terms of pathways through the matrix. However, it is appropriate now to consider how the development of the concepts so far is linked to conventional systems and then to continue developing the approach. Hence, in the next Chapter, systems understanding of rock engineering will be discussed, considering both the implications of what has been developed and the linkages with traditional systems analysis. In Chapter 8, the concept of energy flows will be explored further and the concept of entropy increase as an inevitable consequence of engineering will be introduced. The last two chapters, Chapter 9 and Chapter 10, include discussion of implementation of the systems approach. Systems thinking in rock engineering leads to all kinds of advantages in terms of engineering control, parameter alteration techniques, assessing candidate schemes and systems auditing of rock engineering projects.

7

Systems Understanding of Rock Engineering

Icon 7: Parallel and Serial Nature of Rock Engineering

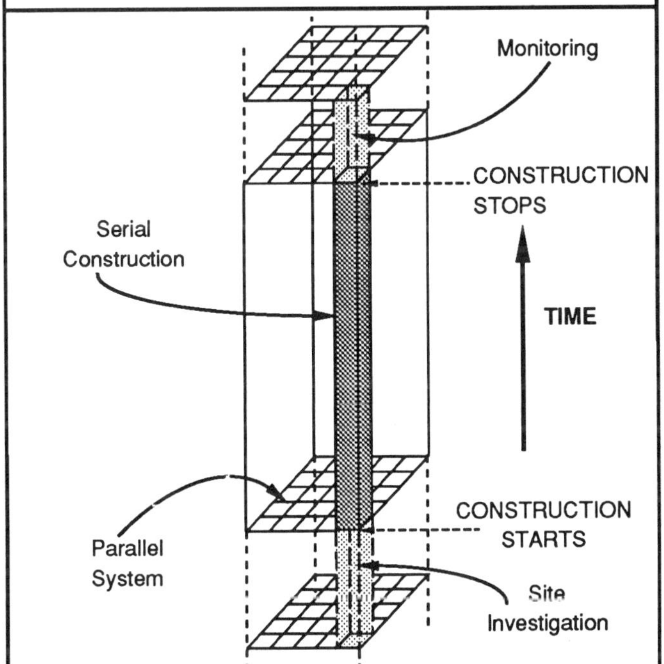

The interaction matrix represents mechanisms operating in parallel. It changes with time both naturally and as a result of rock engineering processes implemented serially.

At this stage in the book, the principle of the interaction matrix has been introduced, the atlas of rock engineering mechanisms has been presented, and the associated theory relating to matrix coding and the cause *vs.* effect diagram have been explained. The method of representing 'multiple' mechanisms has also been discussed. All these have been original developments generated by the author. It is appropriate now to consider any links that may exist between the concepts described here and conventional systems theory. This will enable us to reinterpret rock engineering procedures in the light of both perspectives.

In this Chapter, therefore, five main topics will be considered relating to the systems understanding of rock engineering: the systems overview of rock engineering, system behaviour in general, system boundaries, the systems interpretation of existing procedures, and establishing parameter values.

7.1 SYSTEMS OVERVIEW

The Icon on the previous page illustrates the stage we have now reached in systems understanding of rock engineering. There are parameters; there are mechanisms; and there are energy flows. These are represented by the matrix at the very bottom of the iconic diagram where the mechanisms are operating in parallel. All the rock masses that will be subjected to engineering are currently undergoing the natural processes inherent in this lower matrix.

At some stage, the rock is selected to host an engineering project. A site investigation is conducted, and then, as indicated on the diagram, construction starts. This construction is conveniently represented as the last leading diagonal parameter, and hence is located at the lower right hand corner of the matrix.

Although the processes represented in the matrix are operating together, i.e. 'in parallel', engineering must occur through time and hence the individual construction operations will mainly occur 'in series'. Construction begins with the Cutting of The First Sod and terminates at some stage when the structure is completed. The construction must be mainly serial in nature because time is serial. Thus, the matrix is moved upwards through the diagram as construction proceeds. This means that changes are successively introduced into the leading diagonal construction box and the matrix responds accordingly. All kinds of engineering implications are inherent in this observation — because we wish the matrix to respond in a certain way so that a target matrix is achieved at the end of construction. After construction, the condition of the matrix may be monitored in terms of the way the lower right-hand box is changing. Corrective action may be initiated as required.

Thus, the matrix and all the mechanisms that it represents move through time as indicated by Icon 7 to provide us with a systems overview of all rock engineering. Also, it is to be expected that this systems interpretation is bound to be linked in some way with conventional systems. As shown in Fig. 7.1, there are four main types of conventional systems — as described in the book on the philosophy, analysis and control of evironmental systems by R J Bennett and R J Chorley, "Environmental Systems" published by Methuen in 1978.

These four types of system are as defined as follows. A *morphological* system is comprised of the components and form of the system, and hence this refers to the structure of the interaction matrix — the leading diagonal terms and the mechanisms identified via the off-diagonal boxes, and the possibility of x-fold mechanisms interpreted as pathways through the matrix. A *cascading* system consists of the

Conventional Systems and the Interaction Matrix

There are four main types of system
in conventional systems theory.
All four are utilized via the interaction matrix

1. **Morphological System:**
   The components and form
   i.e. the parameters (the P_i)
   and matrix structure
2. **Cascading System:**
   Dynamic inter-related
   procedures, i.e. the matrix
   is 'switched on'
3. **Process-Response System:**
   Coherent processes and
   responses, i.e. pathways
   through the matrix
4. **Intelligent Control System:**
   Adding construction to perturb
   the matrix in a predictable way
   to achieve the objective(s)

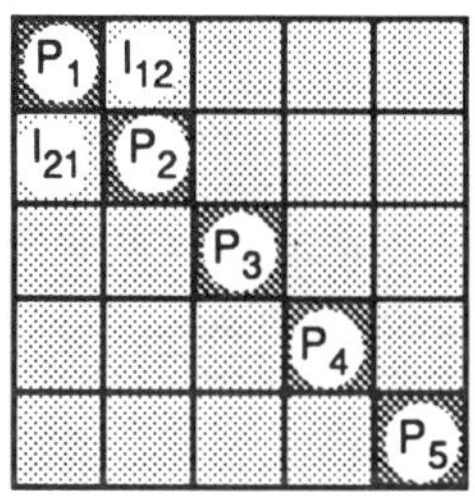

Fig. 7.1 The four types of 'conventional' system used in environmental systems analysis, all of which can be represented in matrix form.

dynamic inter-related procedures, referring in this context to 'switching on' the matrix. A *process–response* system refers to the matrix processes and their overall responses in terms of the leading diagonal parameter changes. The process–response system will therefore include the pathways through the matrix and the matrix reaction to any alteration of the leading diagonal parameters. Finally, there is an *intelligent control* system, which we interpret here to mean that we are adding the leading diagonal construction box to perturb the matrix in a predictable way to achieve the engineering objectives. The matrix interpretation of these four types of system is indicated in Fig. 7.1 and will be developed in the text ahead.

7.2 SYSTEM BEHAVIOUR

Obviously, many of the conventional systems procedures might assist us in developing a systems approach to all rock engineering. However, it is most important at the outset to establish the system most appropriate to the unique nature of the processes in rock engineering. Indeed, it is possible that the direct use of existing systems theories used in other areas might be a diversion, given that a rock mass is a particularly idiosyncratic environment. Hence, the systems approach will continue

The Modes of Matrix Behaviour*

Many different types of systems behave in a similar way, i.e. there is some isomorphism in the laws of science. This is because the same basic rules are governing the relations between the parameters. For example, if the rate of change of each parameter is a function of all other parameters, we will have the series of equations

$$dP_1/dt = f_1(P_1, P_2, \ldots, P_i, \ldots, P_n)$$
$$dP_2/dt = f_2(P_1, P_2, \ldots, P_i, \ldots, P_n)$$
$$\ldots\ldots \qquad \ldots\ldots\ldots\ldots$$
$$dP_n/dt = f_n(P_1, P_2, \ldots, P_i, \ldots, P_n)$$

These equations are true for all such systems, and there are many ways of providing solutions for these to indicate trends in system behaviour

For one parameter dominating

1. A first approximation is $dP/dt = a_1 P$ which has the solution $P = P_0 e^{a_1 t}$
   If a is positive, this is exponential growth;
   if a is negative, this is exponential decay
2. A second approximation is $dP/dt = a_1 P + a_{11} P^2$ which has the solution
   $P = a_1 C e^{a_1 t} / (1 - a_{11} C e^{a_1 t})$

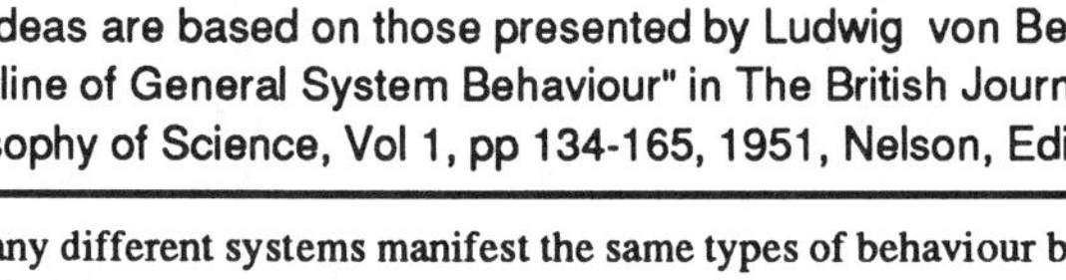

These three curves alone are representative of a great many examples of observed rock mass behaviour

*These ideas are based on those presented by Ludwig von Bertalanffy in "An Outline of General System Behaviour" in The British Journal for the Philosophy of Science, Vol 1, pp 134-165, 1951, Nelson, Edinburgh

Fig. 7.2 Many different systems manifest the same types of behaviour because the basic underlying laws are the same.

to be developed in the first instance here without direct reference to conventional systems. When this has been achieved, it will be appropriate to consider which aspects of existing techniques that have been developed in other fields can be utilized in the rock mass context.

Concepts that are certainly relevant are those that apply to all systems. An example is shown in Fig. 7.2, where the modes of system behaviour (and hence the modes of matrix behaviour) are noted. Many systems behave in a similar way because the same basic rules govern the relations between the parameters. The content of Fig. 7.2 has been extracted from the article "An Outline of General System Behaviour" by L. von Bertalanffy and is very significant for rock engineering. This article points out that if parameter changes can be represented by partial differential equations, then we can expect systems to behave in predictable ways.

As illustrated in Fig. 7.2, if the rate of change of a parameter is directly related to the value of the parameter, then the parameter will change according to an exponential growth or an exponential decay trend and, indeed, there are a large number of cases in rock mechanics where such behaviour is exhibited. This mode of behaviour is not, therefore, some unique feature of rock engineering but represents the underlying nature of the scientific laws. With second order and higher level partial differential equations, the behaviour becomes more complex, but still, of course, following the fundamental mathematical basis. Readers with rock engineering site measurement experience will probably be familiar with all three of the behavioural modes illustrated in Fig. 7.2.

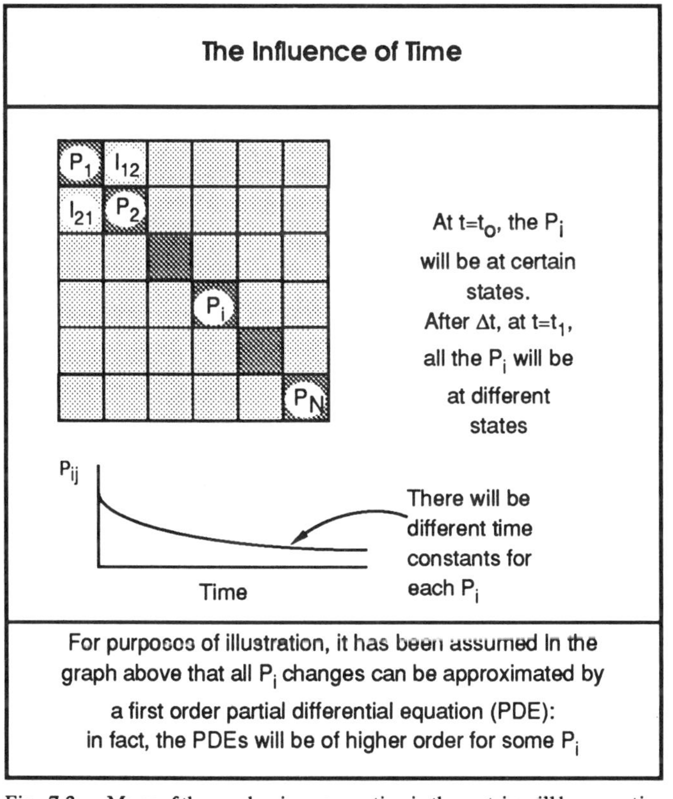

Fig. 7.3 Many of the mechanisms operating in the matrix will be operating at different rates and the results will be superimposed.

One of the possible rock engineering systems approaches, as indicated by the fourth matrix coding method in Fig. 4.3, would be to represent all the mechanisms in the matrix by partial differential equations, either as direct representations of the behaviour or as approximations to that behaviour. A problem in decoding the behaviour would then be encountered as the matrix moved through time. This is indicated in Fig. 7.3 in which it is noted that all the off-diagonal terms are likely to have different time constants, i.e. to be operating at different rates.

Thus, even if it were sensible to approximate all mechanisms by a negative exponential curve, the decay constant would be different for each mechanism. Whether the mechanisms are operating concurrently or consecutively, there is an

Fig. 7.4 Concrete gravity dam in Scotland which showed cracking at the apex.

extremely difficult 'decoding' problem once the matrix has been set into motion — and this would be true even if we actually knew all the mechanisms and the associated characteristic equations.

Consider the photograph in Fig. 7.4 which is of the Mullardoch Dam in Scotland. This concrete gravity dam was built in the late 1950s and was found in the 1980s to be suffering from high stresses at the apex, the cause of which was not known. When the photograph was taken, the water level had been lowered because cracking had occurred in the dam in the region of the white patch at the apex. It was difficult to establish the cause of the fractures in the concrete. A multitude of mechanisms could be invoked to explain such cracking: the geological movement of very large blocks in the region; the effect of thermal fatigue daily from the sun and annually from the seasons; alteration of the concrete properties; and changes in the coupling between the dam foundation and the rock beneath. This is an example of the situation where, after a period of time, it may not be possible to decode the operations of the mechanisms from the final result — the cracking.

A simpler case is illustrated in Fig. 7.5, showing a brick-lined Victorian sewer tunnel (below a city in the UK) which has been subjected to traffic loading from above, a condition not anticipated by the original designers. In this example, the mechanisms are perhaps simpler and the collapse evident in the photograph more readily understood.

However, if there is any general ismorphism in the laws of science leading to systems behaving in similar ways, as illustrated in Fig. 7.2, it should certainly be

Fig. 7.5 Modern day traffic causing failure of old Victorian tunnel beneath UK city.

taken into account. Unfortunately, once the rock mechanics matrix has been set in motion, after the serial construction shown in the Icon at the beginning of the Chapter, and after the structure has responded to subsequent perturbations, we may be unable to decode the observed responses, even though the basic nature of the system is understood.

7.3 SYSTEM BOUNDARIES

Three categories of system defined by the boundary conditions are 'Open', 'Closed', and 'Isolated' systems. (These categories can be applied to the four main system types described in Fig. 7.1.) The categories are defined according to their import and export of energy and mass. In the matrix context and as shown in Fig. 7.6, these system categories can be considered as components of the matrix, travelling from the top left to the bottom right, along the leading diagonal.

For convenience, the energy transfer has been considered in the x-direction and the mass transfer in the y-direction. The definitions of the three categories of system are given in the lower part of Fig. 7.6. They become progressively stricter

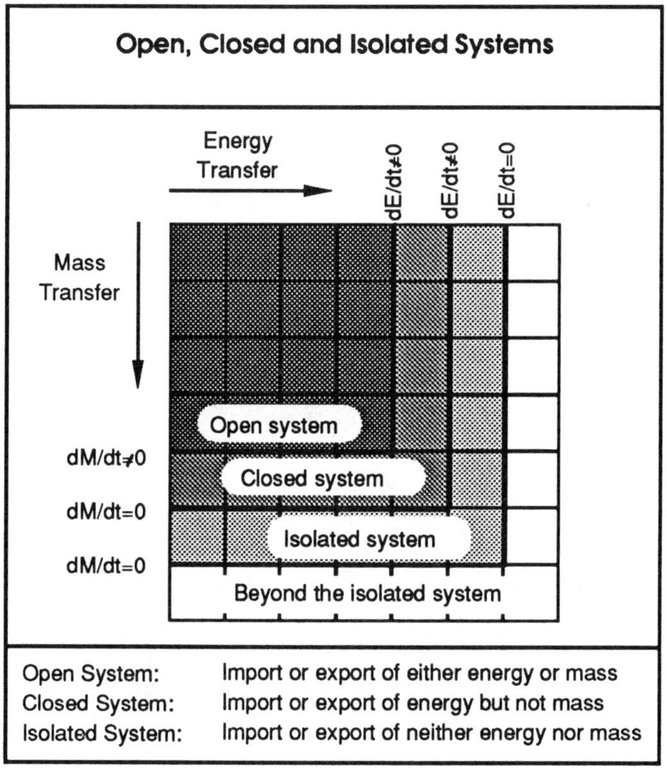

Fig. 7.6 The concept of open, closed and isolated systems in the interaction matrix context.

in changing from an open system through a closed system to an isolated system. The boundaries are located by the rates of change of energy and mass, i.e. where these change from non-zero to zero values. It may or may not be useful to utilize these definitions for the general rock engineering systems approach; they will certainly be useful for specific projects involving water and heat, such as geothermal energy production and radioactive waste disposal.

The more conventional boundary terms used in rock engineering are the 'near field' and the 'far field'. The definition of the near field is that region around the engineering works where the perturbations caused by engineering are greater than any specified change, say 3% as an example. The far field is beyond this near field region, and hence the perturbations in the far field are less than the specified change.

The near field and the far field can also be represented in the matrix, as shown in Fig. 7.7. In this Figure, the rock mass–site–project ultra-coarse 3x3 matrix is shown, with the two extra leading diagonal terms being the near field and the far field. In the engineering sense, all the interactions associated with the far field, i.e.

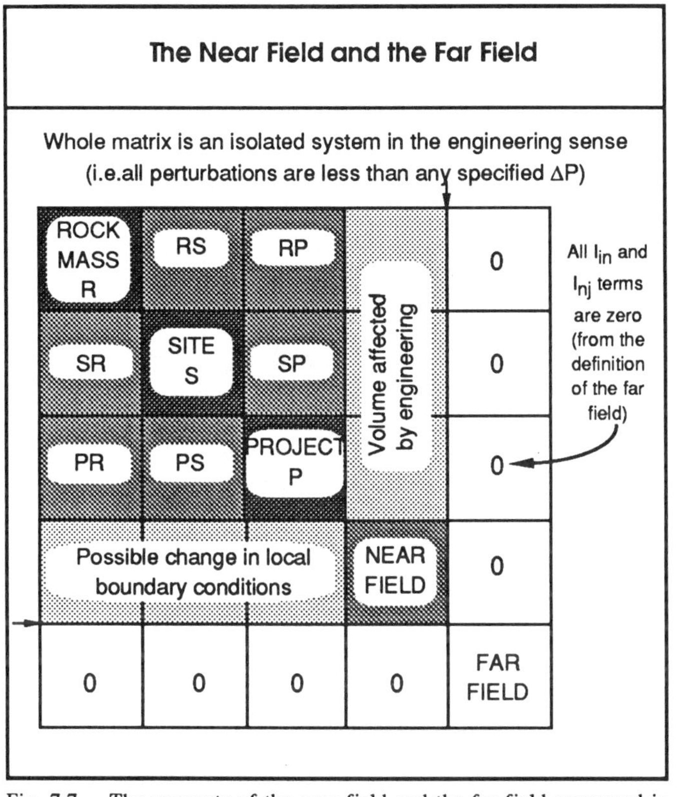

Fig. 7.7 The concepts of the near field and the far field expressed in the matrix context.

the lowest row and right hand column, are by definition zero. Note the small arrow in Fig. 7.7 on both sides of the matrix representing the perturbation value defining the boundary between the near field and the far field.

The way in which the near field and far field concepts are important will depend on the project. In Fig. 7.8 there is a photograph of the well heads at the geothermal project in the Carnmenellis granite in Cornwall, UK. In this experimental project, two boreholes were drilled to 2 km below the earth's surface. Cold water was pumped down one borehole and hot water extracted from the other borehole. The heat reservoir at a depth of 2 km below the surface provided the heat to increase the temperature of the water as it flowed through the fractures in the rock from one

Fig. 7.8 Hot dry rock geothermal site in the Carnmenellis granite, Cornwall, UK.

borehole to another. Naturally, for such an operation to be sustainable, heat must be flowing into the reservoir region at a sufficient rate to continue heating the rock and hence the water. In terms of Fig. 7.7, this circumstance might be rather difficult to represent, although a combination of Figs. 7.6 and 7.7 would provide a very useful representation of the circumstances there.

Most cases are much simpler. When constructing a tunnel, the near field and the far field are usually well defined, although the exercise of establishing precisely what Figs. 7.6 and 7.7 represent could be extremely helpful, e.g. where the ground water level was being lowered.

7.4 EXISTING ROCK ENGINEERING PROCEDURES

Fig. 7.9 is an integration of the Icon at the beginning of the Chapter, and the near and far fields of Fig. 7.7. This diagram represents the systems understanding of rock engineering. It is not only the 'core' matrix (together with the expanded matrix once construction is added) but also the enlarged matrix which includes the near field and the far field that are developing in parallel and changing with time. One can look at the quarry face in Fig. 7.10, and think about the natural process–response system occurring, man's influence on the system, where the near and far fields are, and the fact that this quarrying process will have a beginning and an end.

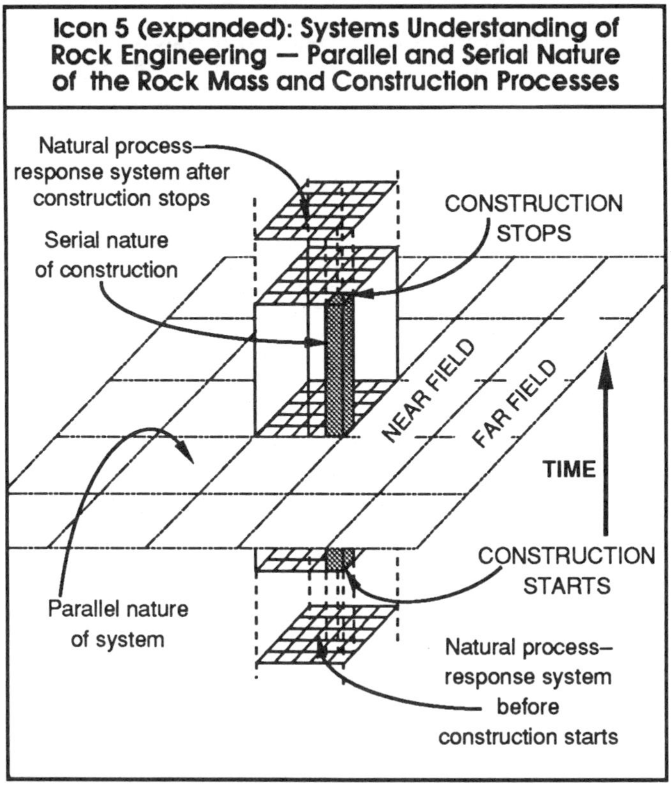

Fig. 7.9 Evolution of the matrix through time as construction continues, from site investigation to project completion.

Fig. 7.10 Limestone quarry in the north of England showing striking discontinuity sets which have an effect on blasting — see cover photo of book.

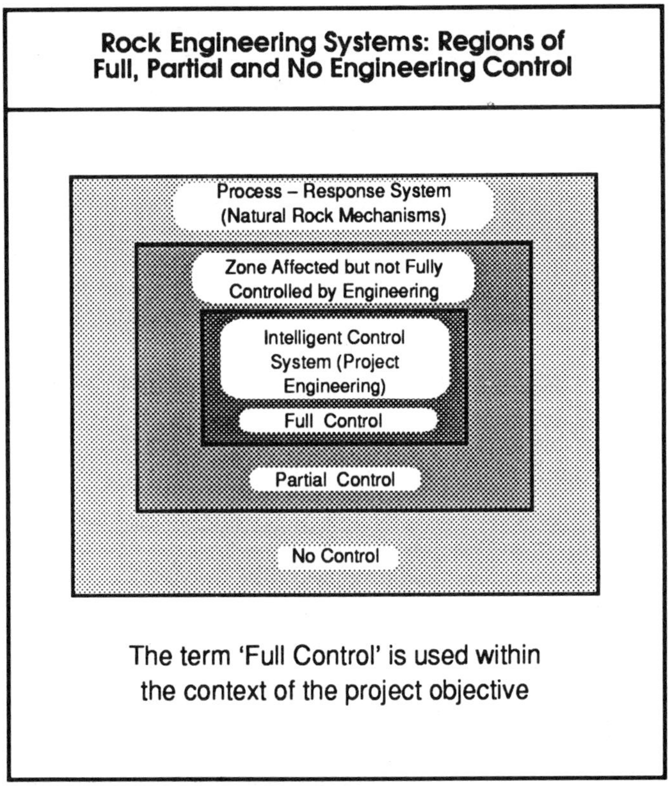

Fig. 7.11 Around the 'controlled' region of engineering there will be an affected region and beyond that an unaffected region.

Fig. 7.12 A quarrying operation is a system — the one illustrated above is in the Mountsorrel granodiorite near Leicester, UK.

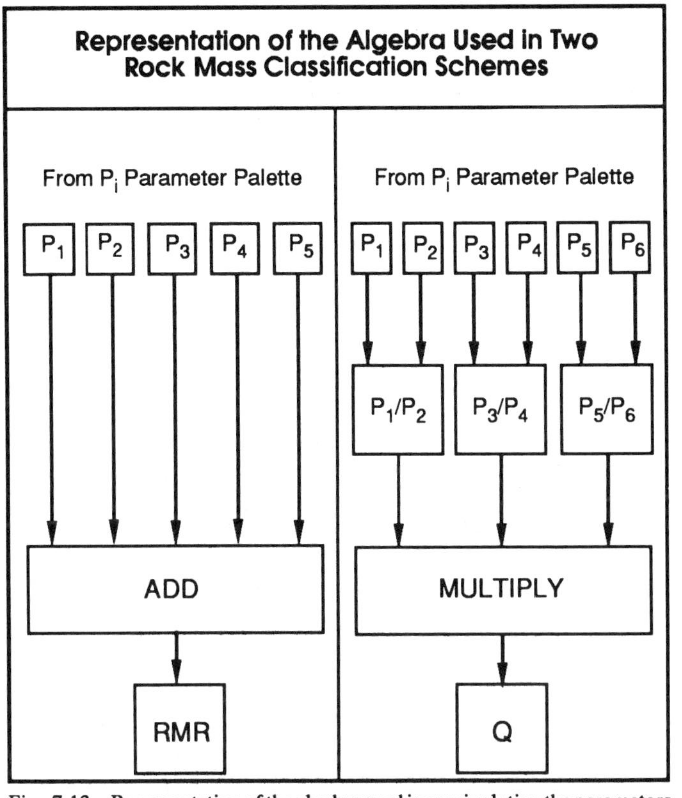

Fig. 7.13 Representation of the algebra used in manipulating the parameters in the two main types of rock mass classification schemes.

In Fig. 7.11, the idea of the intelligent control system is interpreted as project engineering, where in theory we have full control. Outside the central box in the Figure, there is bound to be a zone which is affected, but not fully controlled, by engineering. It could be the rock structure, or the rock stresses, or the water flow, or more subtle aspects of an environmental nature. Beyond this affected zone, we revert to the natural process–response system which is unaffected by the engineering project.

Thus, the overall conceptual understanding illustrated in Fig. 7.9 can be linked with the regions of full, partial and no engineering control in Fig. 7.11. In Fig. 7.9, there is a natural process-response system in the rock before construction starts. Construction perturbs the matrix and the near field. Beyond the near field, the far field continues to operate as a natural process–response system in the same way as the core matrix operated before construction started. Thus, the diagram would repeat because the original core matrix would continue to be applicable in the far field area, but for the purposes of our analysis the system is considered to be completely shown in Fig. 7.9: once the far field is defined, we do not by definition have to consider it.

Another quarry in the UK, this time in the Mountsorrel granodiorite, is shown in Fig. 7.12. There are many ways to extract an aggregate material due to the latitude afforded by the uniform three-dimensional extent of the material. The systems approach can be used to great advantage in these circumstances, not only as a basic checklist to establish unacceptable combinations of parameters, but also to optimize the engineering and to provide imaginative and new methods of extraction.

The discussion in this Chapter has considered the system in a very general sense. Let us now look more explicitly at existing rock engineering procedures and particularly the mechanical characterization of rock.

In rock engineering, there is a need to establish the rock quality at a particular site and, given the complexities already described in this book, many engineers have opted to utilise a number of key parameters and develop a rock mass

Fig. 7.14 The complex nature of a chalk mass, illustrating the futility of attempting to characterize this rock fully.

classification based on the site values of these parameters. The algebra of the two main methods, the RMR and Q methods, is shown in Fig. 7.13. (Both methods are described in "Engineering Rock Mass Classifications" by Z. T. Bieniawski, Wiley, 251 pp., 1989.) For the RMR method, five parameters are used and their values added, with some additional amendments to account for the circumstances. For the

Q method, six parameters are used, quotients taken of successive pairs of the values and these multiplied as illustrated schematically in Fig. 7.13.

These rock mass classification methods for describing the quality of a rock mass have their supporters and their detractors. The supporters say that the methods are very rapid and reliable in providing a quick path through the jungle of potential complexities already described. The detractors say that the sophistication of the

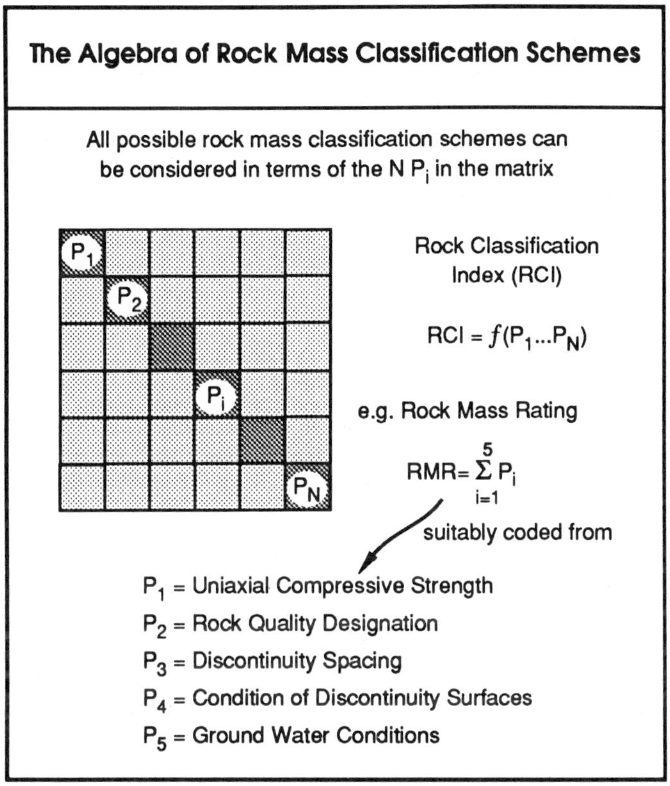

Fig. 7.15 Generic rock mass classification as any function of the leading diagonal parameters of the interaction matrix.

rock engineering mechanisms cannot possibly be represented by a single number and that the presence of, for example, a fault near a tunnel would completely invalidate the approach. Studying the chalk mass in Fig. 7.14, one can see how impractical it would be to attempt to describe the detailed geometry of the fracturing in this rock mass — and indeed the exercise is most likely to be unneccessary. The rock mass classification approach is certainly attractive when dealing with a material such as that illustrated.

However, when approaching the subject of rock mass classification via the systems perspective, there is no difficulty in forming an opinion on the validity of such schemes. Basically, it is necessary to tailor the parameters and the algebra of the rock mass classification schemes to the project in hand. This is represented by the diagram in Fig. 7.15 which is a generic illustration showing that all possible rock mass classification schemes can be characterized simply as a function of the leading diagonal parameter values of the matrix. Note that the two suites of P_i parameters are different for the two schemes. The RMR method is simply the addition of five potential leading diagonal terms of a matrix. The Q method is simply the product of three quotients using six potential leading diagonal terms of a matrix. Naturally, the number of leading diagonal terms and the algebraic manipulation of the coded values can be adjusted according to the project. In some cases, the use of the RMR and Q values will be sufficient; in other cases, it will not be.

The systems approach enables an evaluation to be made based on our knowledge of which mechanisms, and hence which parameters, are crucial in the design, construction and monitoring procedures. We will return to this theme later, when referring to Figs. 9.6a and 9.6b, and suggest a specific method of rock mass classification based on parameter interaction intensity and dominance. Certainly, the systems approach puts all rock mass classification schemes into the correct perspective.

7.5 ESTABLISHING PARAMETER VALUES

There are other equally important ramifications once the systems approach has been adopted. Even if we do adopt a rock mass quality indicator of some type via a rock mass classification scheme, there has to be some assurance that the parameters are being correctly measured. Consider the case of the *in situ* stress. (This is not included in the RMR scheme but is considered in the Q scheme.) We may well wish to include the *in situ* stress state in a tailor-made rock mass classification scheme for a new project. But is the assumed *in situ* state of stress correct?

The perils of ignoring the systems interactions are indicated in Fig. 7.16 for the example of measuring the *in situ* state of stress in a rock mass. The *in situ* stress state might be a leading diagonal term, but can it be measured — bearing in mind that it may be interactive with most of the other parameters? The method of measuring the stress and the associated interpretation may contain the implicit or explicit assumption that the rock is continuous, homogeneous, isotropic and linearly elastic; the rock mass may be none of these.

In Fig. 7.16, there are six errors listed that could affect the estimation of stress at a point in a rock mass. The first four (discontinuities, inhomogeneity, anisotropy and time) refer to inappropriate assumptions. The other two errors may be introduced by the presence of water and by the measurement technique itself. One can easily see just how difficult it might be to establish the state of stress if these errors are occurring through the interactions in the matrix. Indeed, this point has major implications for our choice of the leading diagonal parameters and to what extent we would wish to identify a basic matrix with minimal interactions.

The ideas can be extended to testing methods, as illustrated in Fig. 7.17. If a property of a rock being measured is a leading diagonal parameter in a very fine resolution matrix, i.e. a matrix containing a very large number of explicit parameters, the measurement is likely to be successful because the data are decodable. As shown in Fig. 7.17, an example of a 'simple' test is a uniaxial compression test on an intact rock. This would be one of the mechanisms represented by the off-diagonal terms in the fine resolution matrix on the right hand side of Fig. 7.17. Conversely, in a coarser resolution matrix, we might be attempting to measure the

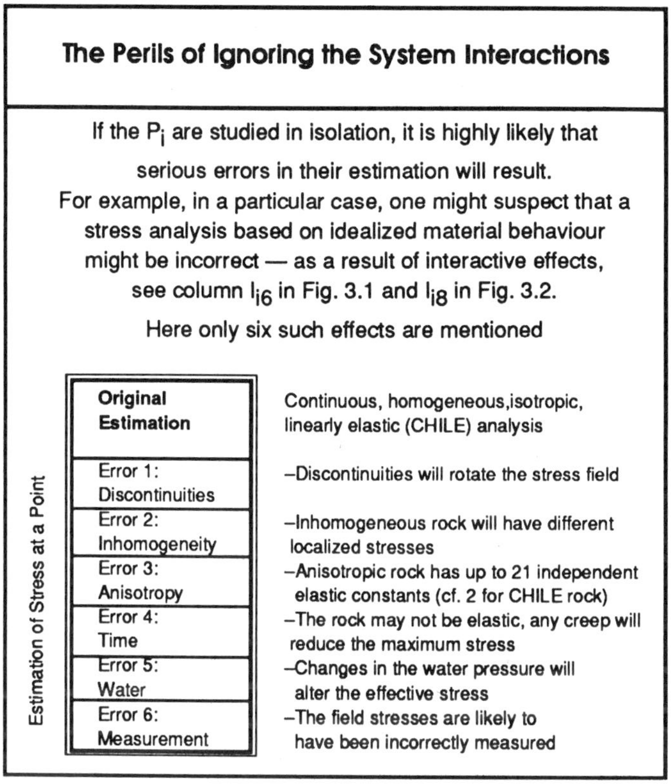

Fig. 7.16 Because many parameters are affecting many others, the process of measuring rock properties is fraught with problems.

in situ modulus of a rock mass (the left-hand side of Fig. 7.17), with all its attendant complications. In this latter case, the data obtained from the test are highly likely to be undecodable and non-transferable.

There are clear implications here for the standardization of rock testing methods. Is the test being conducted in the context of a very fine resolution matrix? If not and if the property is associated with a coarser resolution matrix, it is essential that the complexity of the mechanisms involved in the test be understood.

In Figs. 7.18, 7.19, and 7.20 there are illustrations of the measurement of stress, intact rock strength and discontinuities. These represent three of the most fundamental measurements we can make on a site to assess the mechanical environment within which the engineering will be conducted. The stress represents a mechanical boundary condition, the intact rock represents the fundamental nature of the composition of the rock mass, and the discontinuities determine the rock mass structure.

In Fig. 7.18, the stress is about to be measured by the hydraulic fracturing method in which a portion of a borehole is isolated by inflating two rubber packers, and the portion between the packers is pressurized with water until the rock breaks. In the picture in Fig. 7.18, the lower packer is visible at the bottom of the picture with the mandrel connecting the straddle packer system being the main vertical element in the picture. The difficulty with measuring stress has been discussed

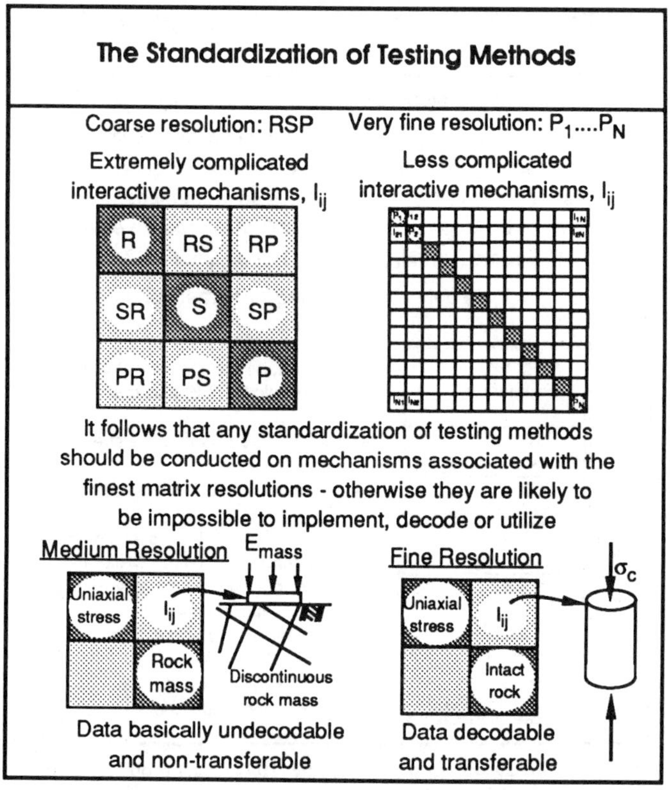

Fig. 7.17 Measurements of rock properties must be 'decodable' — and this depends on the resolution of the interaction matrix to which they refer.

(Fig. 7.16) and certainly most of the errors described have to be understood for the hydraulic fracturing method of stress measurement to be valid.

In Fig. 7.19, the point load apparatus is shown. Using this method of compressing an irregular specimen between two conical platens, it is possible to obtain an estimate of the compressive strength of the rock. This is an index test in the sense that the compressive strength is not being measured directly, but is being inferred from the load at failure. There will naturally be errors inherent in this test similar to those listed for stress in Fig. 7.16. Indeed, all six of the system interactions leading to the errors listed in Fig. 7.16 could apply to the point load testing in Fig.7.19. One should not conclude, however, that there is anything wrong with the test — but simply that its place within the rock engineering system and the interactions inherent in its use must be understood.

Fig. 7.18 Installation of hydraulic fracturing equipment in borehole (mandrel in centre of picture).

Fig. 7.19 Point load test as a good example of an index test that is very simple to use and, most importantly, does not produce results that are difficult to decode.

In Fig. 7.20, the traces of the discontinuities are being measured with a digital compass. The three-dimensional geometry and the mechanical properties of the discontinuities are extensive and complex subjects which have been summarized recently by Professor S. D. Priest in his book "Discontinuity Analysis for Rock Engineering", published by Chapman & Hall in 1992. Clearly the rock structure visible in Fig. 7.20 will require several parameters for even minimal engineering characterization. The level of characterization will depend on both the engineering objective and the nature of the rock.

The three Figs. 7.18 to 7.20 have been presented to show the difficulty of even choosing the rock parameters, let alone actually measuring them on site. The author is particularly concerned about this subject because of responsibility for the International Society for Rock Mechanics Commission on Testing Methods. This Commission produces "Suggested Methods" which are documents prepared by Working Groups to give guidance on the measurement of different rock properties and rock behaviour. At what level should this guidance be given?

Fig. 7.20 Measuring discontinuities on the Isle of Lewis, Outer Hebrides, Scotland.

Despite finding a number of disturbing features relating to the complexity of the rock characterization within the systems context, it is possible to provide the framework for "the way ahead". The rock mass circumstances are indeed extremely complex. For example, *in situ* rock stress is a difficult subject in itself. Stress is a property at a point and yet it is best measured over as large a volume as possible. Similarly, the intact rock material is natural and highly variable over the site, requiring many measurements. The discontinuities dictate the modulus and strength of the rock mass, together with its permeability and local *in situ* stress magnitudes and orientations. But how should the manifold characterisitcs of the three-dimensional rock structure be approached?

Fig. 7.21 Discontinuities in a granodiorite core, illustrating the difficulty in deciding what is and what is not a discontinuity.

Without a structured methodology for assessing the 'level' at which these parameters and the associated mechanisms should be considered and measured, it is impossible to have a coherent approach. The engineering objective must be established. A 'first pass' study must be made of the level at which the approach is to be conducted. All the various stages in the systems methodology must be traversed. Without the systems approach, one can make an intelligent guess, but one cannot be confident that the circumstances have been systematically studied and the optimal engineering inferred.

In Fig. 7.21 there is a photograph of a granodiorite core showing three of approximately one thousand discontinuities encountered in a 417 m length of vertical borehole. The top and bottom of the core are delineated by discontinuities 481 and 483. Discontinuity 482, across the palm of Dr Almenara's hand, contains epidote infilling and appears to be part of the same discontinuity set. Should it be categorized separately? What about the other structural features evident in the core? There are other infilled discontinuities and there are also incipient discontinuities in the core. Should these be recorded? The answer will depend on whether the nature of the engineering project requires a knowledge of these structural features.

Therefore, there can be no pre-determined and fixed site investigation procedures which will apply to all rock engineering projects. The site investigation must be tailored according to the rock, the site and the project utilizing the systems methodology.

In Fig. 7.22, there is a very similar diagram to that in Fig. 7.17 (the earlier diagram illustrated the interpretability of a testing method within the context of the matrix resolution). In Fig. 7.22, the same philosophy applies to the very end of the construction process when we are either monitoring or back analyzing. The monitoring of parameters and interactions in the finest level matrix will be the easiest to understand. Therefore, the decodability of any monitored composite

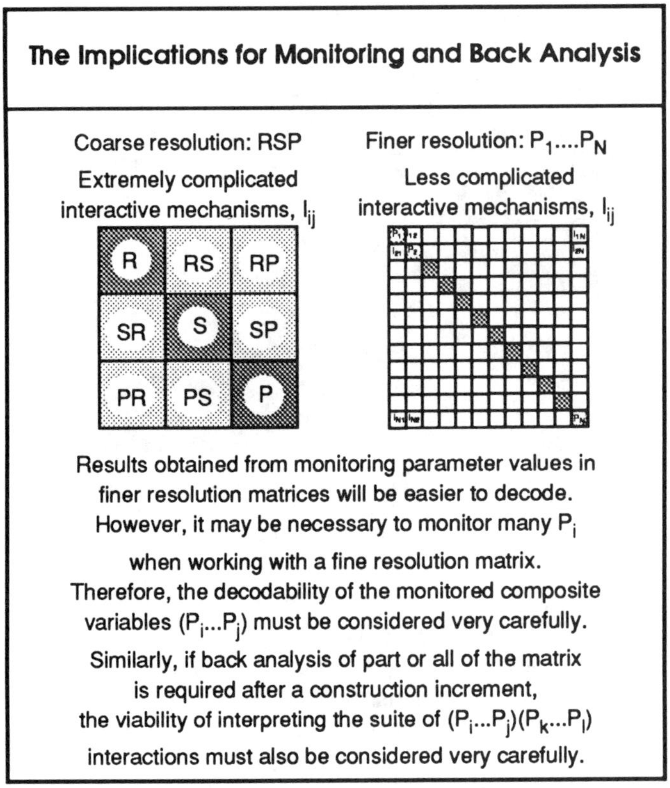

Fig. 7.22 The ease or difficulty of understanding values measured for back analysis will depend on the complexity of the matrix level.

variables must be considered very carefully. In fact, it is through this systems approach that one should actually decide on what to measure, either for quality assurance of the project and/or for any back analysis requirements.

In this Chapter, we have extended the development of the systems methodology described in Chapters 1-6 to incorporate some of the factors involved in the current methods of rock engineering. Much, much more could be said about all the rock engineering procedures in use today and their interpretation in terms of the systems methodology but the examples and discussion that have been presented in this Chapter have been sufficient to illustrate the generic nature of the total problem which, in turn, has implications for the systems methodology itself. We must ensure that the methodology is based on an impeccable foundation, that we understand the differences between the primary state variables, the interactions in the off-diagonal boxes, the energy flows that occur within the matrix, and the evolution of the matrix with time.

Therefore, the next Chapter will be devoted to energy and entropy. These are two of the most fundamental aspects of all rock engineering as interpreted via the systems approach.

8

Energy and Entropy

Icon 8: Energy Flux in the Matrix

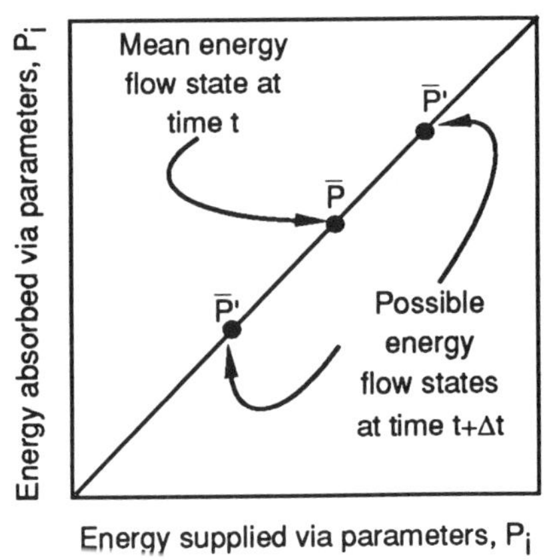

Expressing all the leading diagonal and off-diagonal terms of the matrix in energy units, ML^2T^{-2}, the cause and effect co-ordinates indicate whether the P_i are supplying or absorbing energy. The total matrix energy flux is also indicated.

In this Chapter, we will consider how the approach developed so far can be extended to the general consideration of energy flows within the matrix, corresponding to the energy flows within the rock engineering mechanisms and the project engineering in total. The analysis of energy has proved to be a very useful method for approaching many problems in physics and mechanics. The Griffith theory, for example, considers how a pre-existing crack might propagate when the released strain energy is sufficient to provide the necessary surface energy for the extension of the crack.

One of the main advantages of an energy analysis is that it involves integration of conditions through the whole rock mass; thus the analysis will be more representative of overall conditions than point properties. This is demonstrated by rock strength. When the tensile strength of rock is measured by different testing methods or with different specimen volumes, there are wide variations in the tensile strength values for repeated tests using the same testing configurations. Moreover, the mean tensile strengths determined from tests when the specimen volume is changed or the test conditions are changed (e.g. from the straight-pull test to the beam test) are different. This is because the tensile strength depends on the weakest link in the test specimen — or where the (variable) local stress first reaches the (variable) local strength. This value will depend significantly on micro- and macro-statistical variations within the rock structure. However, the overall conclusion is that the tensile strength depends significantly on the loading conditions of the test and the specimen geometry. A material property does not depend on these. Therefore, the tensile strength is not a material property.

In comparison, the results of a series of elastic modulus tests on intact rock will be much more consistent than test results for the tensile strength. This is because the whole of the specimen contributes to the modulus value. Similarly, the energy absorbed by a specimen in uniaxial elastic loading is given by $\sigma^2/2E$ per unit volume (uniaxial stress, σ ; specimen modulus, E), and it is found during repeated testing that the values of this energy absorbed have relatively little variation.

Thus, it could well be advantageous to establish the energetic interpretation of our systems approach and to consider the implications. The sequence of development will be to see if we can code the interaction matrix in terms of energy levels, then to consider energy sources and sinks in the matrix, to establish the energy flows via the type of pathway interpretation that was established in Chapter 6 and, finally, to consider changes in entropy.

Entropy is a measure of the energy no longer available to drive mechanisms. A change in entropy occurs when energy which is freely available to operate mechanisms is transformed into energy which is diffused and no longer usable. There is, of course, no loss of energy: the increase in entropy is simply a measure of the increasing inability to use some of the energy. Also, increased entropy is associated with greater disorder, with the consequence that all operating mechanisms lead to increased entropy and increased disorder within the *total* system. This must apply to rock mechanisms and rock engineering systems. Therefore, we should consider the entropic consequences of rock engineering.

8.1 ENERGETIC CODING OF THE INTERACTION MATRIX

In Fig. 8.1, we begin the general consideration of coding the matrix for energy and the energy flows associated with matrix mechanism pathways. On the right-hand side of Fig. 8.1, an example coding has utilized integral values from 0 to 4 to illustrate the principle. The choice of integer will depend on the energetic intensity rate (energy flowing per unit time) of the mechanism in the off-diagonal box. It is emphasized that the numbers shown in this diagram are arbitrary and used simply for illustrating the concept.

In this context, the leading diagonal terms cannot be 'conceptual' such as those in the interaction matrices presented in Chapter 3. They must be parameters that can be expressed in energy units. Thus, the leading diagonal terms will have energy values. It is then apparent that the energy-coded value in an off-diagonal

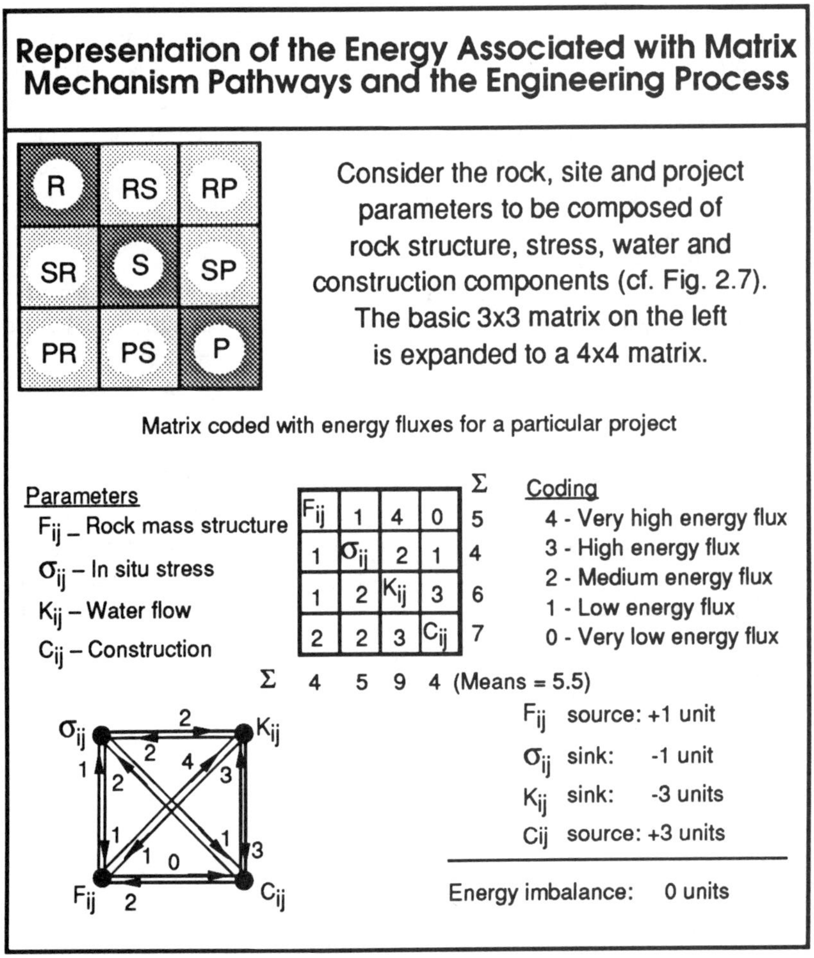

Fig. 8.1 Coding the matrix for energy flows and establishing which parameters are energy sources and which are energy sinks.

box I_{ij} represents the change in the primary state variable P_j induced by the primary state variable P_i .

Hence, the summation of the coding components in a row of the matrix represents adding the energy increments supplied by the parameter source in the relevant leading diagonal box, i.e. how much energy is supplied from that box per unit time. Similarly the summation of the energy elements in a column represents adding the energy increments supplied to the parameter sink in the leading diagonal box, i.e. how much energy is supplied to that box per unit time.

As also indicated in Fig. 8.1, we can introduce the concept of the amount of energy flowing *from the rock mass* and the amount of energy flowing *into the rock mass.* At the bottom left of the diagram, the leading diagonal terms have been represented as a network to indicate the energy flows.

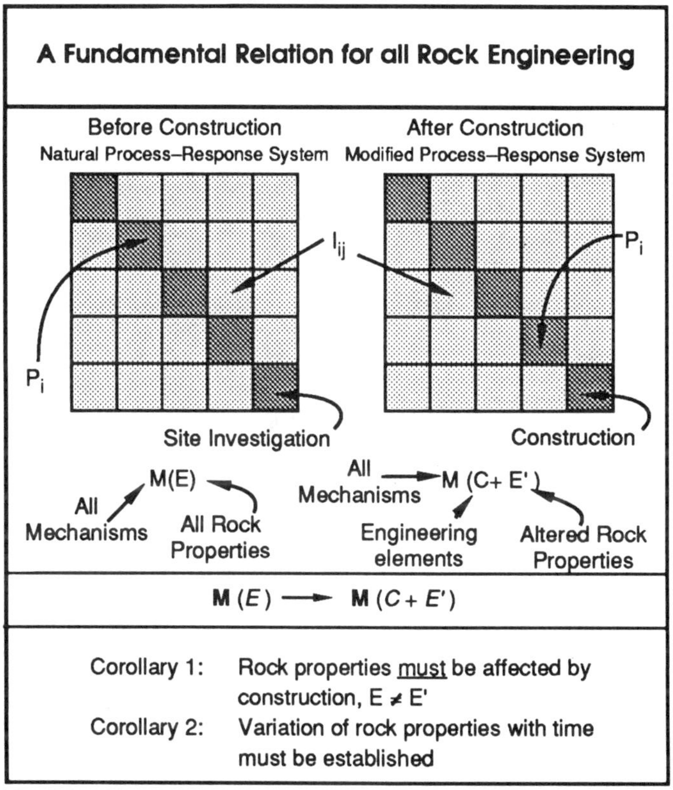

Fig. 8.2 The energy in the total system before and after an increment of matrix operation must be the same — providing a general relation for all rock engineering.

One can now consider a general relation that represents incremental changes in the matrix and hence the overall matrix changes. This is represented in Fig. 8.2 as the fundamental relation of all rock engineering (developed by D. L. Millar, pers. comm. 1991). Assuming that we knew all mechanisms, **M**, operating in the matrix and that we knew all the rock and construction properties, E, we can then consider

M(E), with units of energy, as all mechanisms acting within the rock.

Assume that this is the state of affairs before construction. Now, after construction, the matrix mechanisms, **M**, will remain the same. Thus, utilizing the conservation of energy principle, we arrive at the relation

$$\mathbf{M}(E) \implies \mathbf{M}(C + E') \tag{1}$$

where E are the rock properties, C are any elements introduced by engineering, and E' are the altered rock properties.

Although very general, relation (1) above has extremely important consequences. One corollary of the relation is that the rock properties must be affected by construction and, in general, for the relation to balance, E will not be equal to E'. It follows that the properties established before construction will not be the same as those after construction. We do not yet know whether the alteration will have any engineering significance. The point is that the rock properties measured during a site investigation will not be the ones that govern the rock engineering structure after construction. Perhaps this is one explanation for the ubiquitous 'unforeseen

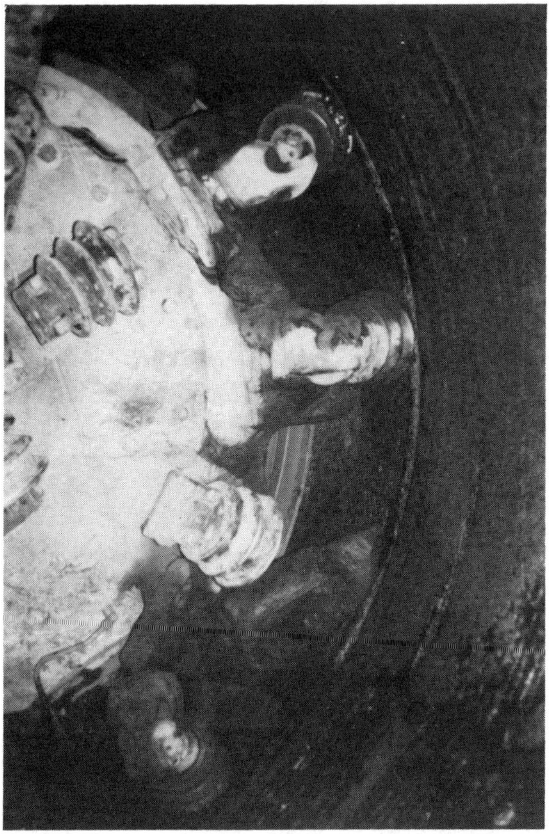

Fig. 8.3 Full-face tunnel boring machine used during the construction of the Kielder tunnel, UK.

ground conditions'! It also leads to a definite recommendation to continue site investigation all through the construction process, and afterwards. It is not the rock's fault that the designer may use the wrong properties, i.e. the ones measured before construction. As the matrix continues inexorably to operate, the rock properties will change with time, both naturally and as a result of construction.

Also, because of the continuing operation of the matrix, the rock properties are not required as fixed values measured at some point in time: we need to know their value as a function of time. Again this may or not be significant. The strength of exposed mudstone changes significantly with time; the strength of unweathered granite does not.

We will never be able to identify all the mechanisms, nor indeed all the changes in rock properties. However, engineers should always be alert to the fact that construction will change the rock properties. There is always the potential for a problem with any rock engineering design that is based on rock properties measured before construction. This potential is therefore not unexpected; its precise manifestation may be.

Consider the action of the full-face Demag tunnel boring machine (shown in Fig. 8.3) during excavation of the 5 m diameter Kielder tunnel constructed to connect the Tyne and Tees rivers in the UK. The process of such excavation interpreted generally is to change the pre-existing rock block size distribution (the rock blocks being formed by the natural discontinuities) to the reduced fragment size distribution (caused by the excavation process). We know that the properties of the excavated rock have changed because of the change in the size distribution — and indeed we could calculate the theoretical energy required to achieve this change. But, have any other rock properties been changed? Does the rock in the tunnel wall (visible in Fig. 8.3) have the same properties as before? Are the deformability properties of the rock the same? Is the strength of the rock the same? Is the permeability of the rock the same? These are the questions that naturally arise from energetic analysis of the interaction matrix.

8.2 ENERGY SOURCES AND SINKS

After the very general considerations illustrated in Figs. 8.1 and 8.2, let us now study the energy analogue to the earlier matrix coding discussed in Chapter 4. The interpretation of matrix energy coding is illustrated in Fig. 8.4. As shown in the Figure, the P_i values represent the energy states associated with each leading diagonal primary state variable and the I_{ij} represent the energy channels through which the changes in the primary state variables occur. As we have already noted, the sum of the terms in a row, which we denoted before as the 'cause', represents all the energy flowing *from a leading diagonal term through the system.* Similarly, summing all the components in a column represents all the energy flowing *from the system into a leading diagonal term,* which previously we termed the 'effect'. It is possible now to consider the energetic interpretation associated with the P_i parameters plotted on the cause *vs.* effect diagram.

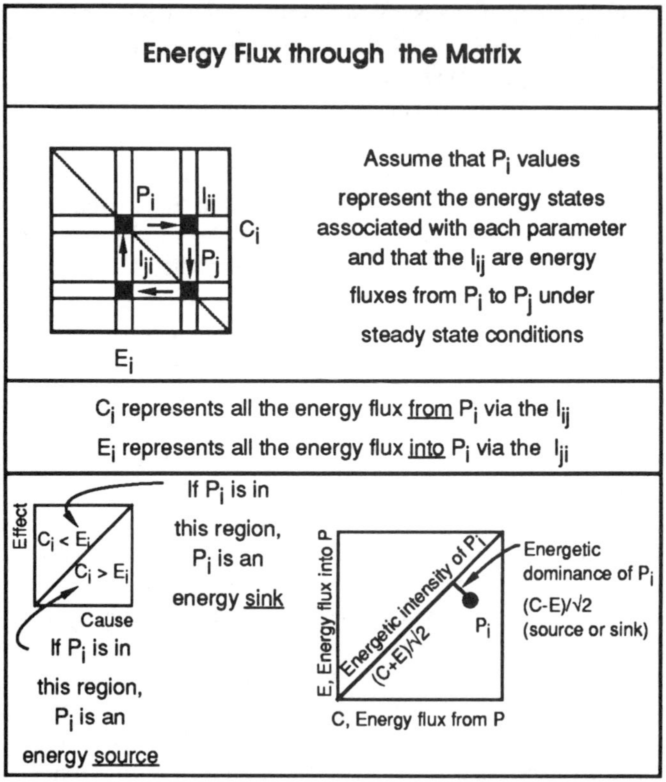

Fig. 8.4 The C *vs.* E plot for an energy coded matrix indicates the energetic interaction intensity and dominance of the P_i .

A parameter plotted on the energetic cause *vs.* effect plot is shown at the bottom right of Fig. 8.4. The energetic intensity of the parameter is represented by the distance along the $C=E$ line; the energetic dominance of the parameter is represented by the perpendicular distance from the $C=E$ line to the parameter. Hence, the energetic intensity represents the flux that occurs via the leading diagonal parameters as they take part in the mechanisms, and the energetic dominance indicates whether a parameter is acting in an overall way as an energy source or an energy sink. If C is greater than E, the parameter is acting as a source; if C is less than E, the parameter is acting as a sink.

The energetic interpretations of Theorems 1 and 2, previously described in Section 4.4, are illustrated in Fig. 8.5. We have the two cases where the cause *vs.* effect plot arises from, firstly, a symmetric interaction matrix and, secondly, an asymmetric interaction matrix. We saw that in the former case the parameter points lie on the $C=E$ line and in the latter case they can lie anywhere in the diagram. However, the mean value of P_i must lie on the $C=E$ line: this is Theorem 1.

For a symmetric matrix, the interpretation of the C *vs.* E plot (shown at the top left of Fig. 8.5) is that the energy input equals the energy output *for each parameter*. The energetic interpretation of Theorem 1 itself is the principle of the conservation of energy: the *mean* source value must equal the *mean* sink value.

For an asymmetric matrix, the interpretation of the *C vs E* plot (shown at the top right of Fig. 8.5) is that the energy input does not equal the energy output for each parameter: some parameters are sources; some parameters are sinks. But, summing over all parameters together, the energy input equals the energy output, and energy is conserved.

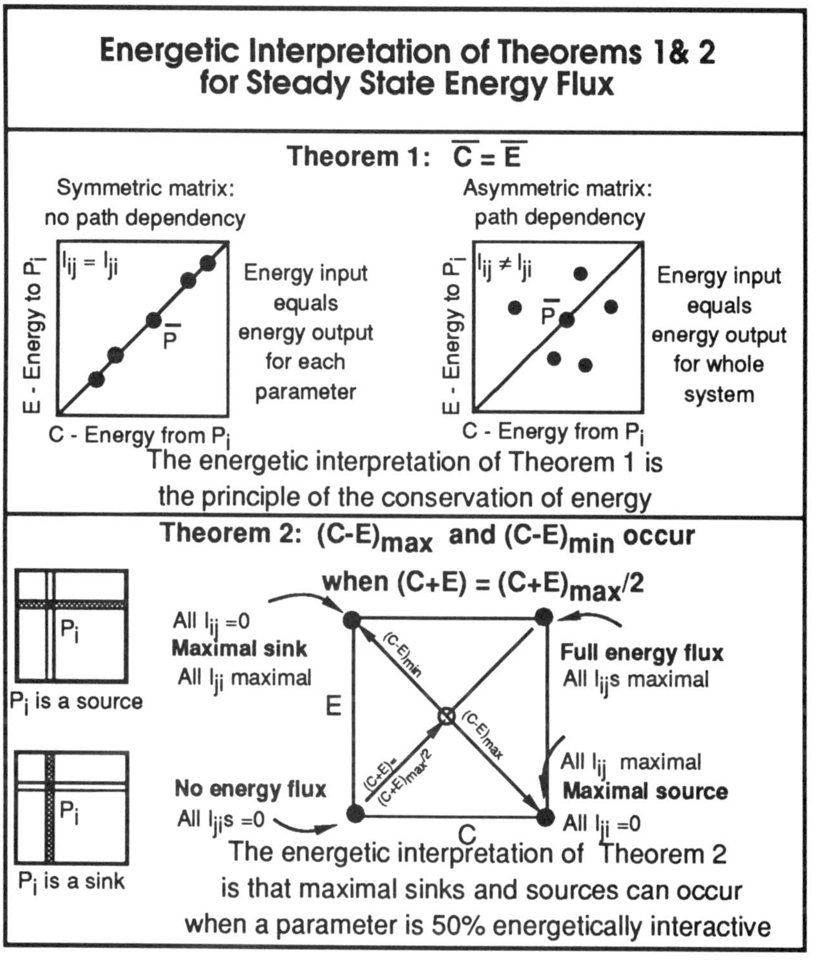

Fig. 8.5 Theorems 1&2 apply when the interaction is coded for energy changes — because the theorems are independent of the coding subject or method.

In the lower part of Fig. 8.5, Theorem 2 is illustrated. Note the four bounding points of the *C vs. E* plot. At the bottom left, $C=E=0$ and there is no energy flux. At the top right, all matrix boxes are operating with maximal energy flux. At the bottom right, for a single parameter, a maximal amount of energy is flowing out, but no energy is flowing in; thus, this is a maximal source for the system. Conversely, at the top left, for a single parameter, no energy is being supplied but a maximal amount of energy is being absorbed; thus, this is a maximal sink. Hence, the energetic interpretation of Theorem 2 is that maximal sources and sinks potentially occur when a parameter is 50% energetically interactive.

The previous discussion supports the idea that we should indeed express the primary state variables along the leading diagonal in energy terms. Calculating the (C, E) co-ordinates is then equivalent to calculating the energy flows in the matrix. Once Construction is added as a leading diagonal term and, bearing in mind that the construction process could either supply energy or withdraw energy (i.e. act as either a source or a sink), then the value of P_N, Construction, could be located anywhere on the C,E diagram. Also, the mean $\bar{P}_i$ could either go up or down the $C=E$ line after a construction increment Δt (see Fig. 8.6).

An example of construction supplying energy is the full-face tunnel boring machine shown in Fig. 8.3. An example of construction withdrawing energy would be the

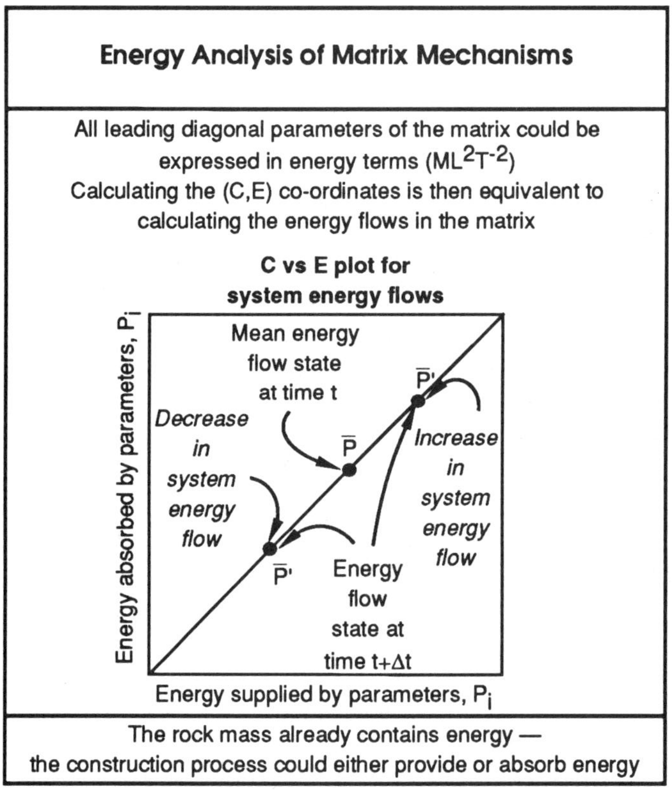

Fig. 8.6 The interactive energy intensity changes after a time increment can be represented via the C *vs.* E diagram.

block caving method of mining, where the orebody is undercut and then the rock mass above withdrawn through draw points. The strain energy of the rock mass is utilized in creating fractures and altering the pre-existing natural rock block size distribution to the mined fragment size distribution. The amount of energy released by the rock is far greater than that used to develop the mining operation.

With all these energy transfers operating, we have to ask the question "Is the engineer in control?" One definition of engineering would be the successful control of the energy flows in the matrix in order to achieve a target matrix. Fig. 8.7 shows how we start with a natural process response system represented by a matrix with dimension N-1. The project engineering or construction is then added at the bottom

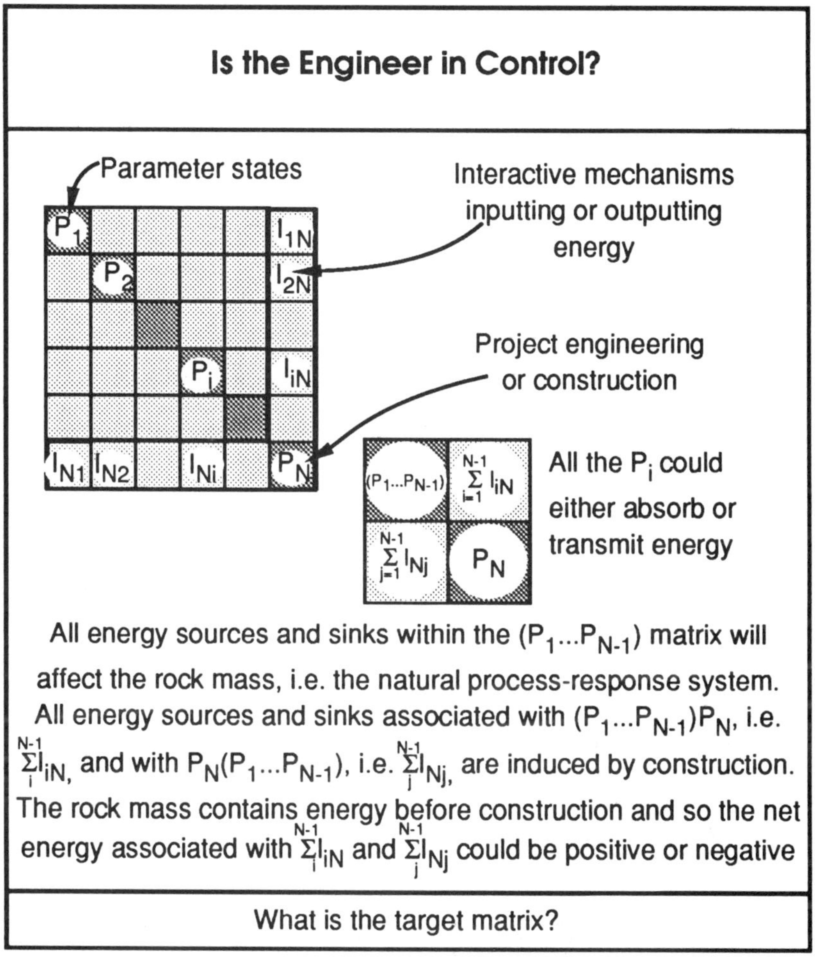

Fig. 8.7 Analysis of the matrix interactions will establish whether the engineer can control the flow of energy during construction.

right as the Nth leading diagonal box. From the associated row and column, we can study the energy being transferred by construction or activated by construction. The net energy associated with the construction process could be either positive or negative. In Fig. 8.8, Case 1 covers circumstances where there is more energy input than output; Case 2 covers circumstances where more energy is output than input. As mentioned before, Case 1 is well illustrated by a civil engineering tunnel; Case 2 is well represented by the block caving method of mining. However, it is

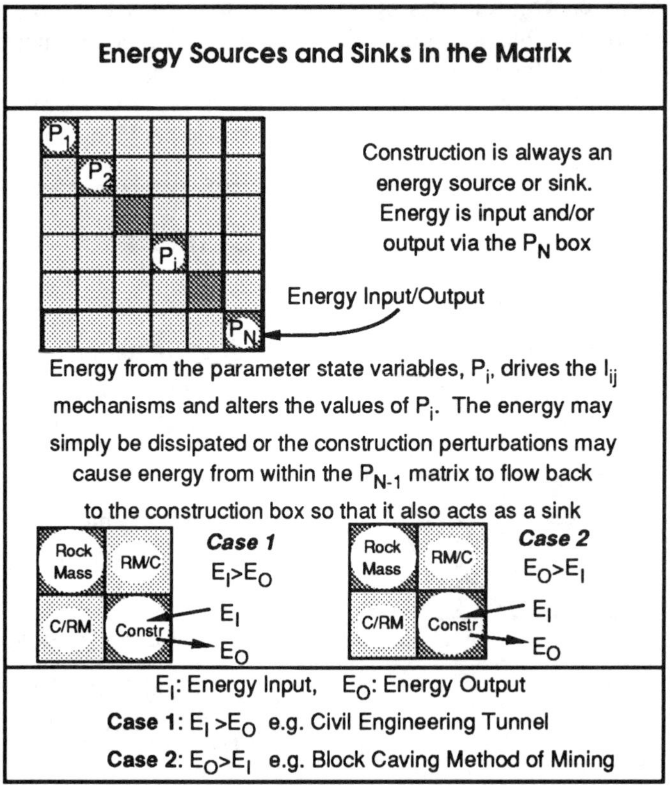

Fig. 8.8 The type of rock engineering activity will determine whether the rock mass is absorbing or releasing energy.

insufficient to consider solely the net energy, either input to the rock mass or output from the rock mass. We need to study all the energy sources and sinks in the matrix — and treat construction as one of these.

In Fig. 8.9, there is a photograph of a core box showing core lengths of intact rock from the Carnmenellis granite in Cornwall, UK. It is in this granite that a geothermal demonstration experiment has been underway for many years on the potential for extracting the natural heat in rock for energy production (see also Fig. 7.8). At 2 km depth, where water is travelling from the injection borehole to the extraction borehole, the maximum component of natural horizontal *in situ* rock stress is about 60 MPa. It does not take much imagination to realise that vast quantities of energy are already stored in the form of elastic strain energy in the rock mass and that the injection of water could severely alter the energy stability of the fractured rock mass at depth. Indeed, one of the features of the geothermal programme has been the very successful identification of water flow and associated block shearing through interpretations of recordings of the microseismic emissions, as described by R. J. Pine and A. S. Batchelor in their paper "Downward migration of shearing in jointed rock during hydraulic injections" (Int. J. Rock Mech. Min. Sci. & Geomech. Abstr., 21, 5, 249-263, 1984).

Fig. 8.9 Core lengths (1.5 m) of the Carnmenellis granite in which the hot dry rock experiments are being conducted, UK.

8.3 ENERGY PATHWAY FLOWS

As rock engineers we are often dealing with rock masses that contain quantities of energy far beyond our comprehension. It is the manipulation of this energy to achieve the target matrix state that dictates the design, construction and monitoring techniques to achieve the project engineering objective.

Thus, the control of the energy in the rock mass via the perturbations that we introduce through construction requires an analysis of the multiple mechanisms represented by pathways through the matrix. This is illustrated in Fig. 8.10. When a perturbation is introduced by 'switching on' the construction, i.e. sending a pulse to all the P_i, many pathways are activated. The P_i energy states can act as sources or sinks and the process will generally be altered so that the matrix again stabilizes, having accommodated the energy perturbation. Exceptionally, the matrix will become locally unstable at a certain time due to a particular combination of the P_i energy sources and sinks and associated I_{ij} mechanisms — after which the matrix will restabilize.

The two cases, stability and instability, are conceptually represented in the lower part of Fig. 8.10. A practical example is the rock slope in Fig. 8.11. What would happen if a block were removed from this slope? There are three main possibilities: nothing would happen; the removal of one block would cause further failure; the

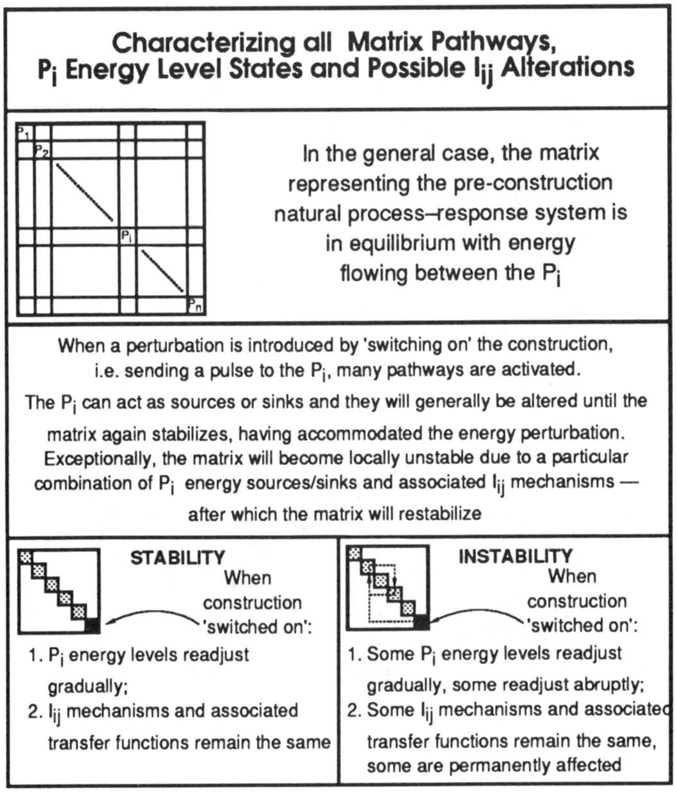

Fig. 8.10 The matrix will stabilize when a perturbation is introduced in the construction box unless an instability pathway is activated.

whole slope would collapse. Which of these is regarded as stability depends on one's interpretation. The 'nothing would happen' scenario is an engineering interpretation of stability. The 'whole slope would collapse' scenario is an energetic interpretation of stability, the rock mass reconfiguring to a more stable state.

However, before considering any further direct engineering interpretations, it is necessary to consider in more detail the concept of energy pathways within the matrix. The development of this subject is potentially huge and only the directions in which the research might take can be indicated here. The subject is strongly connected with some well-established branches of mathematics, such as catastrophe theory, network theory and graph theory. The reader is referred to the books "An Introduction to Catastrophe Theory" by P. T. Saunders, Cambridge University Press, 144pp., 1980, and "Graphs and Networks" by B. Carre, Clarendon Press, 277pp., 1979 for further information on these subjects.

A very old problem in mathematics is the 'Travelling salesman problem': given the position of N towns, which is the shortest route from the salesman's base

Fig. 8.11 Rock slope stability lies in the eye of the beholder.

through each of the towns and back to base? This is illustrated in Fig. 8.12 with an example taken from the book "Decision Mathematics" referenced at the bottom of Fig. 8.12 for five towns around Sheffield in the UK. Given the positions and distances from Sheffield of Aston, Chapeltown, Chesterfield, Rotherham and Worksop, which is the shortest route from Sheffield through each of them back to Sheffield? The network at the top left of Fig. 8.12 is represented in matrix form in the Figure.

There is no closed form solution to the travelling salesman's problem: it has to be solved algorithmically. The connection between this problem and rock engineering is that we are interested in energy pathways which involve either the least or greatest amounts of energy. These might represent engineering extremes such as optimal stability and total collapse. Establishing the associated pathways is similar to the travelling salesman's problem. Consider the example matrix in the lower part of Fig. 8.12. What does the path through the shaded squares represent?

The general concept of 'visiting each P_i once' matrix pathways is shown in Fig. 8.13. When there are only two leading diagonal parameters, there is only one pathway (assuming we start at the lower right-hand matrix component, Construction). When there are three leading diagonal parameters, there are only 2 pathways. When there are four leading diagonal parameters there are 6 such pathways. In the general case, there are $(N\text{-}1)!$ separate pathways connecting each P_i once and only once, given that the starting box is specified.

For a large dimension matrix, we could construct an energy density histogram representing the number of pathways associated with certain class intervals of pathway energy flux — see the lower part of Fig. 8.13. This type of energy pathway analysis would have to be conducted via some form of numerical analysis, but the

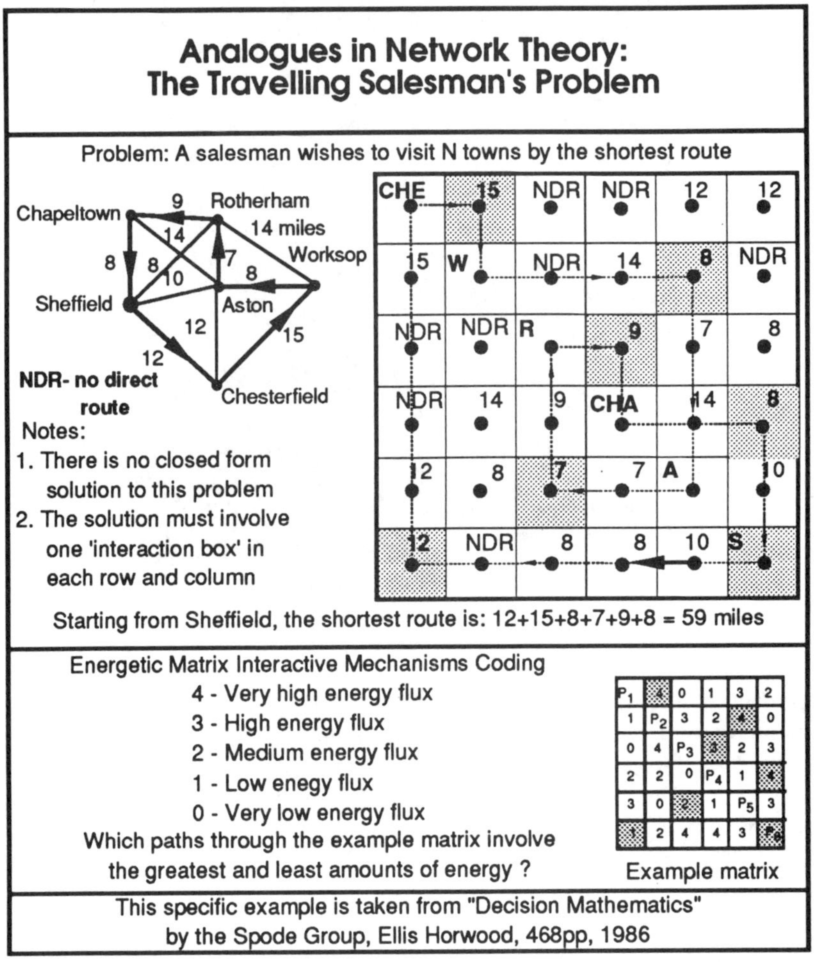

Fig. 8.12 There are analogues between matrix pathway analysis and some traditional mathematics problems.

analysis would be different in principle to finite element, boundary element and distinct element methods of establishing stress and strain distributions because of the pathway aspects.

The consequence of all these energy movements within the interaction matrix is highlighted in Fig. 8.14 in which the ideas of attenuation, stability and entropy are presented together. The horizontal axis in the diagram in the lower central part of the Figure represents the leading diagonal of the interaction matrix. The columns represent the energy levels (or more strictly the energy generation potential or the energy absorption potential) associated with the parameter primary state variables. This diagram again indicates that the *N*th box, Construction, could be an energy source or an energy sink.

To remind us of the possible engineering realization of various states in Fig. 8.14, the two photographs in Figs. 8.15 and 8.16 represent parts of the Seikan undersea tunnel between the islands of Honshu and Hokkaido in Japan. Fig. 8.15 shows an uncontrolled inrush of water during construction, implying that a part of the matrix is 'out of control'. During this period of uncontrolled water entry from the sea

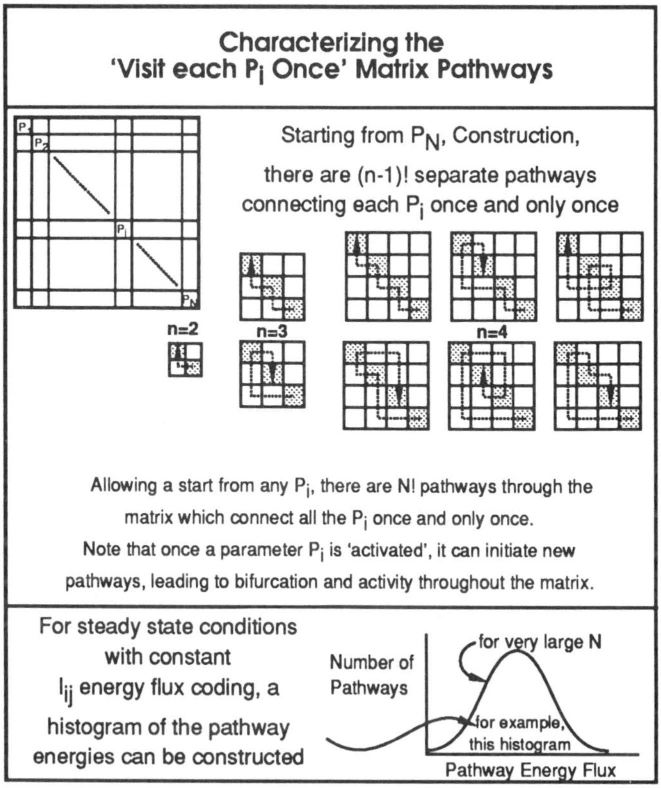

Fig. 8.13 A histogram of matrix pathway energy values can be compiled to study the form of the distribution of pathway energy flux.

above, engineers would not have known whether the situation was recoverable, i.e. whether continued engineering would result in the target matrix being reached. The fact that indeed this was possible is shown by the successful completion of the tunnel shown in Fig. 8.16 — one of the most spectacular civil engineering achievements of all time.

The conclusion so far is that it is going to be essential to conduct energy analysis of the mechanisms and hence the parameter evolution modes, implied in Fig. 8.17. This Figure also refers to the concept of entropy which we will consider next.

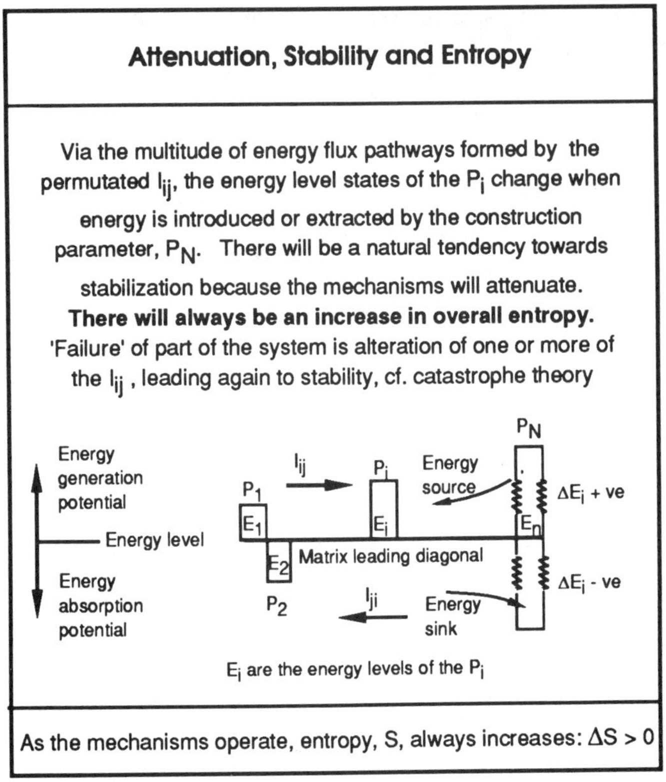

Fig. 8.14 Energy sources and sinks along the leading diagonal of the interaction matrix.

Fig. 8.15 Inrush of water during construction of the undersea Seikan tunnel linking the islands of Honshu and Hokkaido in Japan.

Fig. 8.16 The completed Seikan tunnel, Japan..

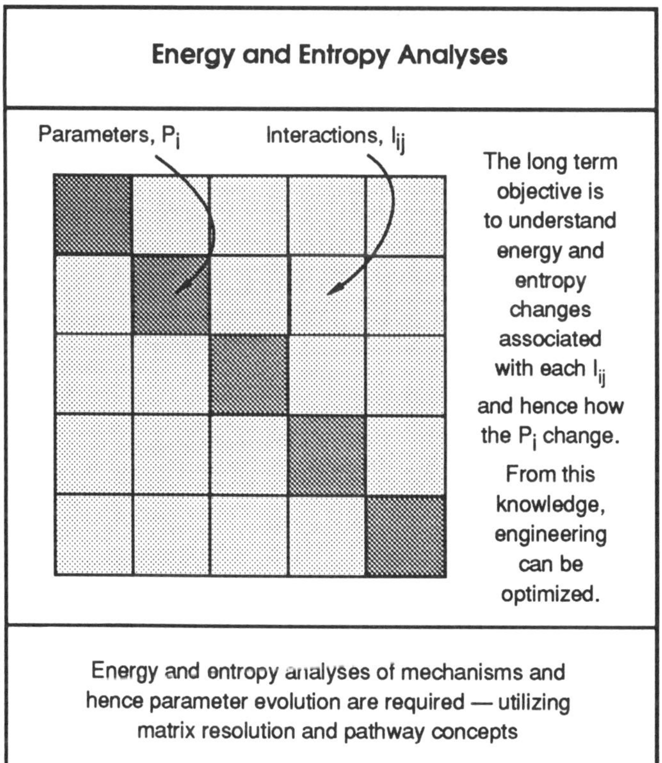

Fig. 8.17 The full analysis of matrix operation requires an energy element and pathway approach.

8.4 ENTROPY

The progression of these ideas has led to the conclusion that the fundamental analysis should be concerned with the energy and entropy resulting from all the mechanisms operating in the matrix. Energetic and entropic analyses of mechanisms should be conducted utilizing the matrix resolution and pathway concepts already discussed. With regard to entropy, it has already been mentioned that the operation of any mechanism will result in some usable energy being dissipated into a non-usable

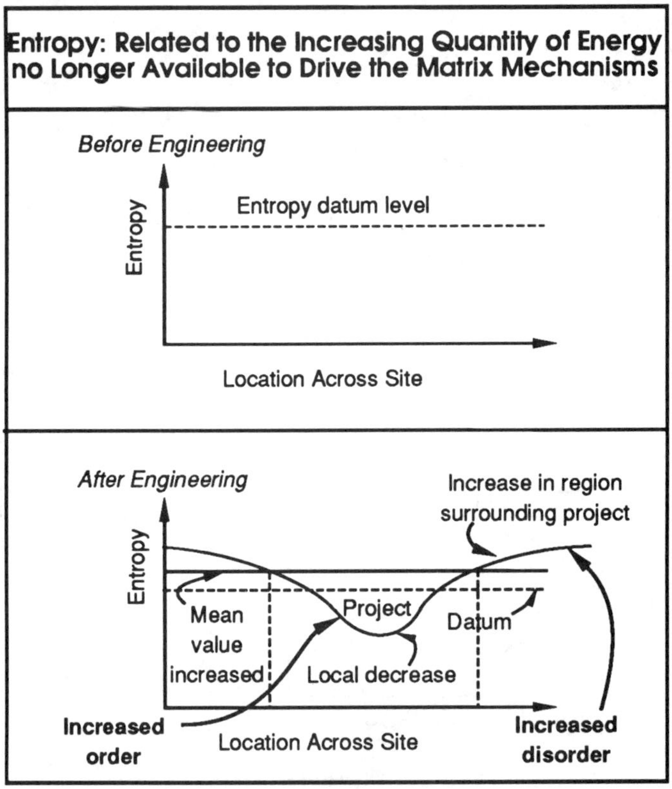

Fig. 8.18 Engineering is associated with a local reduction in entropy but a more than compensating increase in the surroundings.

form. This causes an increase in entropy which is associated with greater disorder within the total system. Readers interested in the wider aspects of this subject are referred to the book "Entropy - A New World View" by J. Rifkin, Bantam Books, 302 pp., 1980.

The implications of the fact that entropy, and hence disorder, must increase when any mechanism operates are enormous. Whenever any rock engineering is conducted, the overall result is greater disorder than order. Obviously, if we build a dam, a slope, a shaft, a tunnel, a cavern, or anything else, some energy will be utilized on 'useful' work and some of that energy will always be dissipated. Cumulative energy dissipation will monotonically increase with construction (all active boxes in the matrix contributing to the process). Entropy will monotonically increase. For any order that is created in the total system, greater disorder will also be created in the total system.

Fig. 8.18 illustrates an engineering project that creates order in some region and greater disorder somewhere else. Assume that before construction we take the entropy to be at a certain datum level across the site, as shown at the top of Fig. 8.18. After engineering, as indicated in the lower part of Fig. 8.18, the average value of entropy is increased, because energy will be dissipated into a non-usable form. There is a local decrease in entropy related to the order created during engineering by achieving the project objective. However, there is a more than compensating increase in entropy beyond the project itself, and a more than compensating increase in disorder.

It cannot be emphasised enough how important this is. The concept has major significance for all of man's activities and the future state of the planet. Anything that we do will result in some form of increased disorder. There are as many examples of this as there are mechanisms operating. To take just one in the current

Fig. 8.19 Order or disorder? To the mine manager, this is order; to the environmentalist, this is disorder.

context, consider the plate from which you ate your last meal. The plate represents a high level of order (compared to the clay from which it was manufactured). It has a specific shape, it is often decorative, it has a coating of glass and it serves a function which all of us appreciate. What is the disorder associated with a plate? Some of it is shown in Fig. 8.19 in a china clay open pit mine — alteration of the countryside.

There are many mechanisms involved in making a plate: the first of these is obtaining the raw material for the plate. To do this we are disturbing part of the environment. As discussed in connection with the definition of slope stability in the previous section, in fact the state of order or disorder in Fig. 8.19 is debatable.

To the mine owners, great order is exhibited here and everything is going to plan. The disorder created by the order of the mine would have to be sought in the surrounding region. We conclude that there is a hierarchy of order and disorder, and there will be an associated entropic hierarchy. This brings us full circle and confirms that we have to know the engineering objective: in fact, we need to know the complete hierarchy of objectives related to any project. After all, without this knowledge, how can we know when we have succeeded? Later we will see that the methodology based on the systems approach in this book is objective-based.

Another conclusion is that one of the most important contributions we can make to the future environment via this rock engineering systems approach is to introduce the concept of the entropic audit. In Fig. 8.20, the entropic audit is conceptually illustrated in the matrix containing the project, the near field and the far field. We

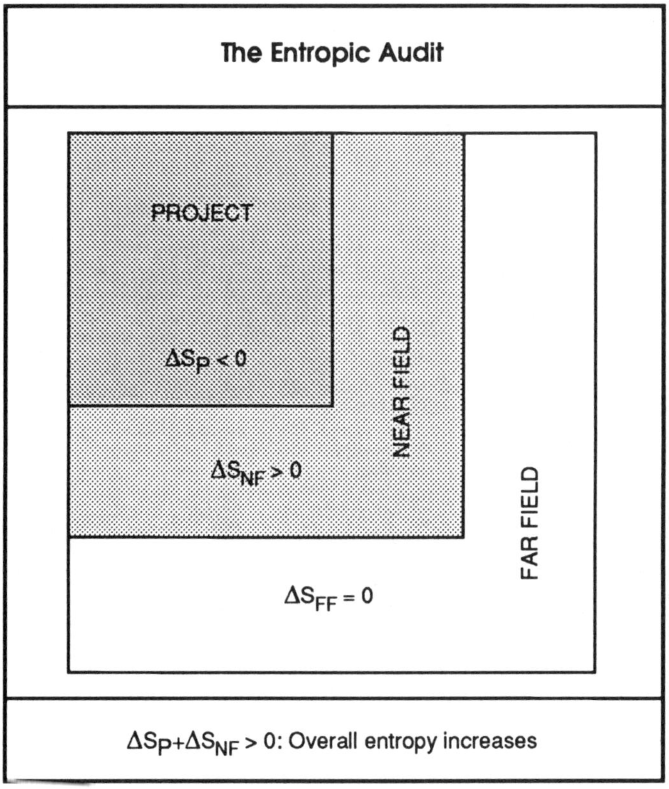

Fig. 8.20 To assess the amount of disorder caused by engineering, an 'entropic audit' must be conducted.

saw in Fig. 8.18 that there will be a local decrease in entropy around the project itself, and a greater increase in entropy in the surrounding area. (This is similar to the development of life forms, which "consume negative entropy", as it has been expressed.) In Fig. 8.20, from the definition of the far field, the change in entropy as a result of the engineering, ΔS_{FF}, will be zero. The change in entropy will be positive for the near field and negative for the surrounding region: the absolute value of ΔS_{NF} will be greater than the absolute value of ΔS_P.

We are all familiar with financial audits. Most people also appreciate the need for environmental audits. It has been shown here that an energetic understanding, i.e. audit, is necessary in order that we, as rock engineers, understand precisely what it is we are doing as we achieve our objectives in civil and mining engineering.

We will discuss later that the recommendation is to have four main types of rock engineering audit: these are the financial audit, the environmental audit, the energetic audit, and the entropic audit. The entropic audit is the highest level of audit and thus very sophisticated. We already know that engineering will create greater disorder than order. The entropic audit establishes what disorder will be associated with the engineering works, and will assist in indicating how to minimize the disorder. Entropy increase is assured; what is required is minimum entropy increase assurance.

This ends the conceptual aspects of the rock engineering systems approach. In the next Chapter methods for implementing these ideas in engineering practice will be discussed. In the last Chapter of the book, we will consider 'systems thinking' in rock engineering and all the attendant practical advantages for design and construction.

9

Implementation of the Systems Approach

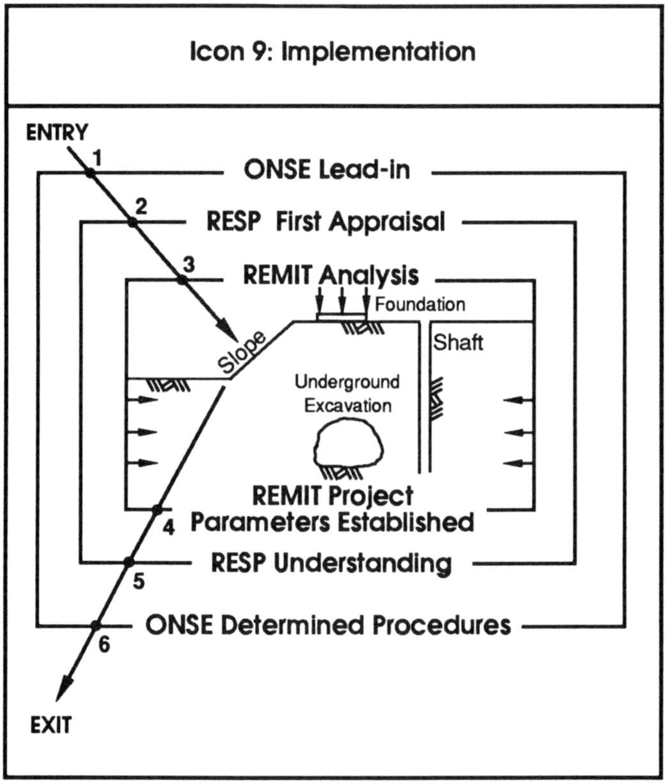

In the implementation phase of the systems methodology there are three 'top-down' analytic steps and three 'bottom-up' synthetic steps. The single most important aspect of implementation is establishing the project objectives.

Following the presentation of the background theory in Chapters 2-6, the discussion on systems understanding of rock engineering in Chapter 7, and in the light of the energy and entropy concepts noted in Chapter 8, the implementation of the systems approach is explained in this Chapter.

It will be recalled that the Icon for Chapter 1 on page 1 represented the two main methods of rock mass modelling: the synthetic model and the analytic model. The advantages of the analytic model were pointed out and we have also explored the consequences of initiating a 'top-down' analytic model in this book. However, once the content of the analytic model has been established, i.e. the model has been mapped out in terms of the relevant components, the process can be reversed and a synthetic model built with these components.

This is the technique which is illustrated in Icon 9 on the previous page. In essence, the system is entered on an analytic basis and exited on a synthetic basis. Three acronyms are used in the Icon. These refer to the REMIT/RESPONSE (or R^2) method of implementation which will be described in Section 9.4:

REMIT is Rock Engineering Mechanisms Information Technology;
RESP is Rock Engineering System Performance; and
ONSE is Objective-based Network Sequence Evaluation.

9.1 THE STRATEGY

The methodology must be embarked upon with an objective. It is not trivial to say that we must know what we are trying to do in order to be able to succeed. As has been seen, the many projects that require rock engineering knowledge and methodologies can have widely different objectives. In civil engineering, the objective is usually the creation of a permanent opening for some function. Along the way, in order to create the opening, some rock will have to be broken. Then, we require the remaining rock to be stable for the design life of the project. For mining, the rock also has to be broken because the objective is to obtain minerals from the rock mass. There is a huge variety of asssociated mining methods — some involving rock stability and some involving rock instability (usually a combination of both). Thus, over the wide spectrum of rock engineering, there will be different objectives associated with excavation and support, both between projects and within the same project, e.g. considerations of the permanent shaft and access tunnels of a mine compared with the the impermanent stopes.

With reference to both the Icon on the previous page and Fig. 9.1, the basic strategy of implementation is via six steps, the first of which is establishing the objectives (Fig. 9.1). The second step is to make a first pass analysis of the rock engineering system performance (RESP) and what might be involved. This is achieved via the study of a generic matrix containing the important leading diagonal parameters. We will be discussing later, with reference to Fig. 9.4, the suite of parameters that might be used for different types of project. Also, the use of the matrix has been the main underlying theme of this book and later, with reference to Fig. 9.5, we will consider the process of tailoring the matrix to the project.

The third step in Fig. 9.1 is then taken: this is the final analytic step, being the

REMIT analysis of specific boxes in the matrix. For example, an explicit block or stress analysis may be conducted. It will be necessary to study to what extent the information necessary for dealing with the off-diagonal boxes in the matrix is actually available. Our knowledge can be enhanced by literature surveys, discussion with experts, study of precedent practice and by utilizing many forms of analysis.

After taking these first three steps, we have now fully 'entered' the problem by establishing the objectives, making the first pass at an initial understanding of the process–response system, and deciding on and indeed conducting some of the specific analyses required for the project. As intimated in the iconic diagram at the head of this Chapter, the method is entirely general: the project could be a foundation, a slope, a shaft, an underground excavation, or any other structure — or indeed any combination of these as required for the project objective.

It is at this stage, rising from the trough in Fig. 9.1, that the methodological mode changes from analytic to synthetic, as we follow the exit path illustrated in Icon 9 and Fig. 9.1. The project parameters have been established and the matrix has been tailored to the project in hand following the associated objectives. We now have the specific knowledge infrastructure for synthesizing the construction procedures. This consolidation and structuring of the knowledge base is the REMIT-specific fourth step.

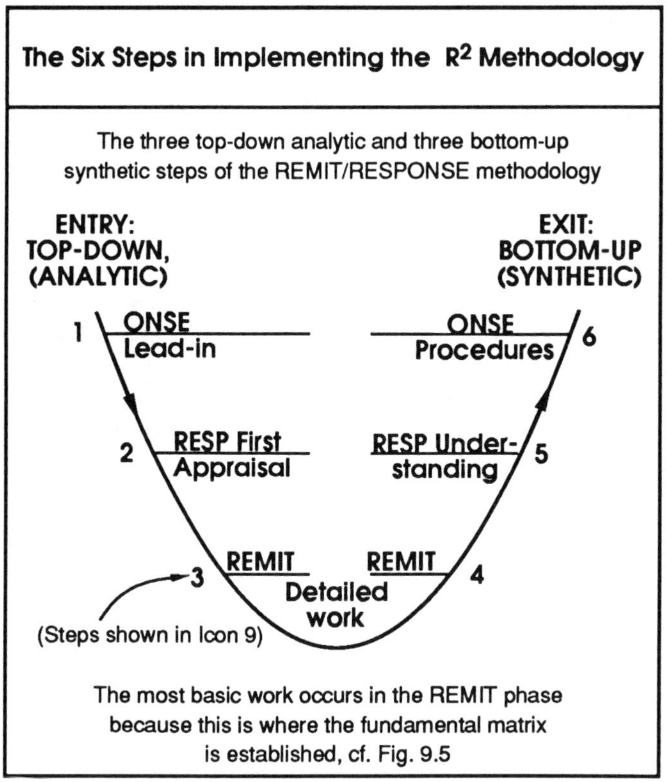

Fig. 9.1 A combination of the analytic and synthetic approaches is used in the implementation of the rock engineering methodology.

The fifth step is the complete understanding (in so far as is possible commensurate with financial and practical constraints) of the process–response system associated with the project and its objectives. In other words, we can 'switch on' our project-specific matrix in order to establish the viability of particular construction options. This is a particularly dynamic stage of the synthesis. We are building an operational procedure which is optimized with respect to the objectives and takes into account the possibility of unacceptable matrix pathways, i.e. disaster scenarios.

Finally, we reach the sixth and last step: ONSE. This step is the establishment of the precise construction design, excavation procedures, support procedures, and

Fig. 9.2 Excavation perations at the Mountsorrel granodiorite quarry, UK.

monitoring. The financial and associated contractual procedures will also be established at this stage. The systems approach will provide guidance on the best financial framework for the project because one can identify the critical links in the construction development.

Of the six steps in implementing the systems approach just described, the first three are analytic and the last three are synthetic. Figs. 9.2 and 9.3 are now included to provide an opportunity to pause and consider the wider aspects of this overall strategic approach to rock engineering.

Fig. 9.2 is another view of the granodiorite quarry at Mountsorrel in the UK, which has an output of 5Mt/y. The objective of this quarry is to provide aggregate — through excavating and crushing the rock. The photograph shows a classic quarrying operation with a dump truck in the foreground travelling down to be loaded by the loader in the right centre of the picture adjacent to the recently blasted bench. Seasoned rock structure observers will note that, despite two main vertical sets of discontinuities, the rock fracturing in this quarry is quite complex. From a rock engineering point of view, 'local' plane, wedge or toppling instability may be acceptable at a bench scale, providing that it does not inhibit safety or operations in any significant way. The key to the rock engineering analysis here is the identification of any major discontinuity planes that traverse the whole quarry and that could lead to a failure of a scale involving several benches similar to the one illustrated in Fig. 1.7.

Taking an underground example, in Fig. 9.3 the junction of the two draft tube tunnels in the surge chamber of the Alto Lindoso hydroelectric scheme in Portugal is shown. Consider how strategically vital the stability of these tunnels is to the operation of the hydroelectric scheme. For the quarry, shown in Fig. 9.3, almost

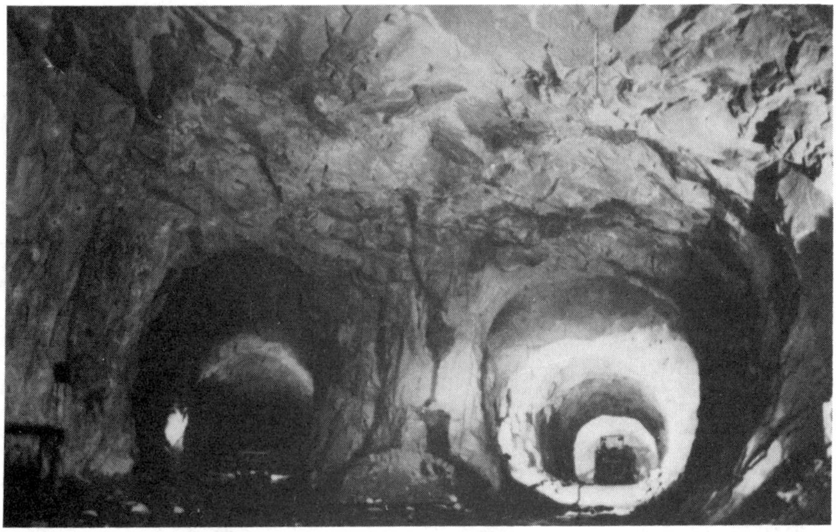

Fig. 9.3 Junction of the two surge chamber tunnels in the Alto Lindoso hydroelectric scheme, Portugal.

any type of surface collapse can be overcome. Imagine, however, the costs associated with the temporary loss of function of the hydroelectric scheme if a major collapse were to occur at the junction shown in Fig. 9.3 once the operation of the hydroelectric scheme had been initiated. (Both the quarry and the tunnels are shown as examples to assist in the thinking process — no problems are actually evisaged with either of these specific rock engineering examples.)

The preceding discussion has highlighted yet again that our objectives can be quite complex. The objective is not simply to build, for example, the rock infrastructure for a hydroelectric project, but also to ensure the long term achievement of this objective, i.e. the maintenance of stability over the design life of the project. It is critical for our understanding of the rock engineering system performance to have anticipated the development of the matrix and all the associated possibilities in Steps 3–4 which are illustrated in Icon 9 and Fig. 9.1.

Naturally, the longer the matrix operates the greater our difficulty in predicting its behaviour. However, it is not acceptable to simply retreat from the problem, saying that the behaviour of the matrix is not predictable over the next 100 years because new variables will be introduced which were not present during construction, e.g. the introduction of high-pressure moving water into an unlined rock tunnel. We must use our engineering imagination to forestall the development of any identifiable adverse scenarios, a theme to be continued in later Sections of this Chapter.

One question that may be asked is "How do I know which parameters should be used as the leading diagonal terms of the matrix?" Having established the objective for the project in the first of the six steps in the systems approach, the Rock Engineering System Performance 'first pass' analysis can utilize 'conventional'

Parameters that could be used for a 10x10 Matrix Evaluation of Three Types of Project

	Water Pressure Tunnels in Hydroelectric Schemes	Large Underground Caverns	Radioactive Waste Repositories	Slopes
ROCK	Rock mass geometry, Discontinuity persistence, Discontinuity aperture, Discontinuity fill	Rock type, Elastic modulus of intact rock, Discontinuity orientation, Discontinuity frequency, Discontinuity aperture, Elastic modulus of rock mass	Porosity, Density, Strength, Elastic modulus of rock mass, Time dependent properties, Discontinuity geometry/ Permeability	Intact rock quality, Discontinuity geometry, Discontinuity mechanical properties, Rock mass properties
SITE	*In situ* stress, Hydrological conditions, Topographic factors, Presence of folds/faults	*In situ* stress, Hydrological conditions, Presence of faults	*In situ* stress, Hydrological conditions	Overall environment, *In situ* stress, Hydrological conditions
PROJECT	Location of tunnel, Water pressure in tunnel	Depth of cavern	Induced displacements, Thermal aspects	Slope location, orientation and height, Support/ Maintenance, Construction technique

These parameters are not comprehensive, but should form the core of any analysis

Fig. 9.4 Parameters of importance for four different rock engineering project objectives. These parameters, which are listed in rock, site and project categories, are examples of the parameters that could form the leading diagonal terms of generic matrices for these projects.

parameters derived from the various sources already discussed. In Fig. 9.4, there is a Table of the rock, site and project parameters that could be used for water pressure tunnels, large underground caverns, radioactive waste repositories, and slopes. This Table is, of course, generic in nature and would form a basis for an initial 10x10 matrix generated by literature survey information and precedent practice.

In the fullness of time, the author intends to develop a single project-independent generic matrix for Step 2 of the systems approach. This matrix will be the unified basis from which tailored matrices can be developed for any project. Until then, the generic matrices presented in Chapter 3 are available for slopes and underground excavations. Other generic matrices are quite easy to produce once the leading diagonals are chosen and after a little practice. In the next Section, the technique of transforming a generic matrix into a project-specific matrix is presented.

9.2 THE TACTICS

The tactics involved in implementing the systems methodology are now described. These are mainly related to identifying which parameters are required, producing the project-specific matrix, and understanding its operation — as we exit through the three synthetic steps in Icon 9.

Readers interested in a specific example of the application of the methodology are referred to the paper "A stability hazard indicator system for slope failure in heterogeneous strata" by C.P. Nathanail, D A Earle and J A Hudson, Proceedings of the EUROCK '92 Symposium, 14-17 Sept. 1992, Chester, UK, Thomas Telford Ltd, London. The R^2 methodology was used for the work reported in this paper in order to establish the slope instablility hazard potential at Ffos Las, a large open-cast coal mine in Wales. The slopes matrix presented in Fig. 3.1 was coded according to the first method listed in Fig. 4.3, i.e. the on-off binary switch method. A very large number of parameters were considered because of the complexity of the site and the many potential slope failure mechanisms. As a result of the application of the R^2 methodology, it was possible to produce a geotechnical proforma that could be completed for various zones in the mine. By these means, the slope instability hazard could be evaluated with confidence, because the proforma had been tailored specifically to the rock, site and project conditions.

In general, one has to establish which is the relevant matrix for the project and then to consider its operation, both as a function of time and as a function of the perturbations introduced by construction. The basic method for establishing this matrix is illustrated in Fig. 9.5. Some of the words in this diagram are new in this context and will be explained first. 'Winnowing' is a word meaning to examine in order to select the desirable elements (and is also used to mean separating the wheat from the chaff). There is an enormous number of rock properties and mechanisms — but which are required, given the project objectives? The term 'hierarchical winnowing' means that rather than individually eliminating parameters from, say, a 300x300 base level matrix, the parameters are eliminated in blocks as the matrix is expanded from a coarse resolution to a fine resolution.

As shown in the left-hand descending portion of Fig. 9.5, as the resolution increases, elements of the matrix are progressively eliminated if they are below the 'importance threshold' applying at that resolution. This is analagous to one of the methods of representing shapes in computer programming, known as 'quadtree' and 'octree' representation: the term quadtree applies to two dimensional shapes; the

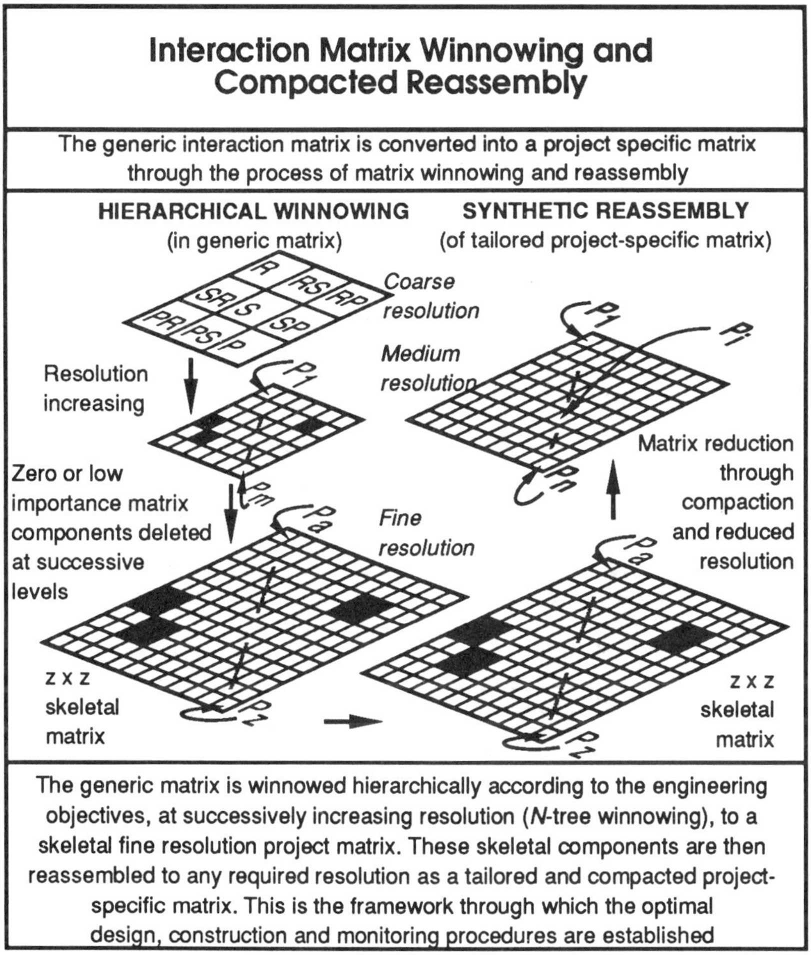

Fig. 9.5 The winnowing process by which a generic interaction matrix is tailored to a project-specific interaction matrix (Steps 3 & 4 in Icon 9 and Fig. 9.1).

term octree applies to three dimensional shapes. For quadtree representation, a square surrounding the shape is divided into four equal sub-squares. Each sub-square is coded on a binary basis, 1 or 0 according to whether the shape in question is in the sub-square or not. Then, each sub-square is divided into four more sub-sub-squares, each of these being similarly coded, and the process continued to the resolution required. This is known as quadtree representation (the 'quad' referring to the four squares and the 'tree' referring to the increasing resolution).

If the shape is three-dimensional, a cube is used to surround the shape and is then divided into eight sub-cubes, each sub-cube coded, and the process repeated over and over again, and hence the term 'octree'. This method of defining the shape of a body is much more efficient and has greater utility than simply specifying all the x, y, z co-ordinates of the shape.

The same method is used here in the hierarchical winnowing in the left hand part of Fig. 9.5. The generic matrix is hierarchically winnowed according to the engineering objectives at successively increasing resolution. With the explanation of the quadtree and octree terms, this process can be termed N-tree winnowing. The end product is a skeletal fine resolution project matrix.

At the top of the left-hand side of Fig. 9.5, obviously we cannot delete anything from the 3x3 coarse resolution matrix with the rock, site and project as leading

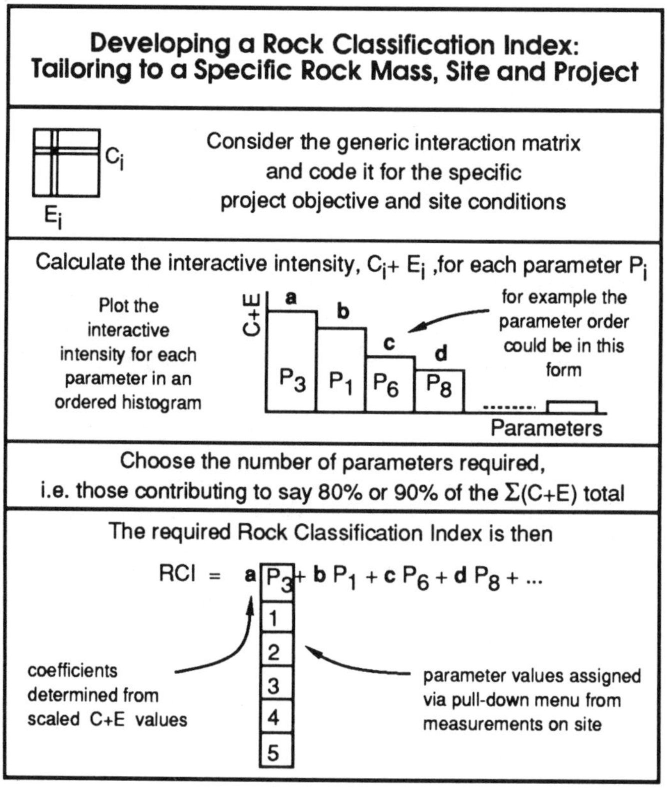

Fig. 9.6a Developing a rock classification index that takes into account the project objective and parameter interaction intensity.

diagonal terms, because they are naturally all above the importance threshold. At intermediate resolution, with leading diagonal components P_1 to P_M, there may be various blocks that can be deleted, because they do not relate to the project objectives. For example, the chemical sorption of particles onto rock fracture surfaces could be deleted *en bloc* if rock slope stability is our objective. After hierarchical winnowing of the generic matrix, we arrive at a $Z \times Z$ fine resolution skeletal matrix — which could be sparse for 'simple objective' type projects. For projects with several functional components and an associated objective hierarchy, the fine resolution ZxZ matrix could be quite dense.

Having established this 'base level matrix', the skeletal components need to be reassembled to any required implementational resolution as a tailored and compacted project-specific matrix. This is the framework from which the optimal design, construction and monitoring procedures are established.

The reader might be interested in the logic of developing a rock classification index tailored to a specific project and site as part of this 'recompaction' process. In other words, how could we improve the existing rock classification index systems were we to be utilizing the methodology proposed in this book. The development of such a classification index is illustrated in Figs. 9.6a and 9.6b.

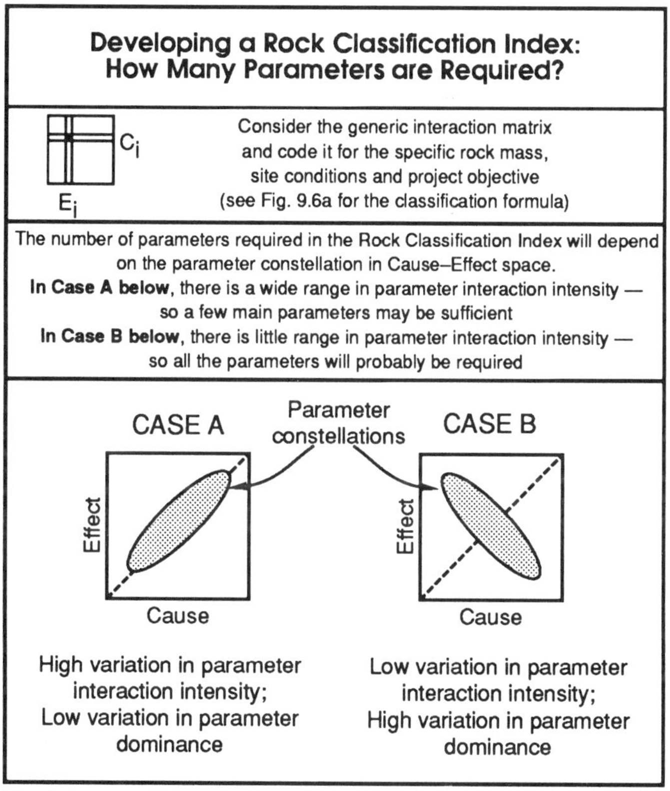

Fig. 9.6b The number of parameters required in a rock mass classification scheme will depend on the form of the *C vs. E* constellation.

We could either start with the generic interaction matrix or the project-specific matrix. As noted at the top of Fig. 9.6a, each term in the matrix would be coded according to the project objectives. Several methods of coding have been discussed in Section 4.1. The coding could be according to the intensity of the interactions, or the actual energy involved in the mechanisms, or according to a more subjective criterion concerning whether the activation of the mechanism in the off-diagonal box is of benefit or disbenefit to the project. A histogram can be compiled according to the $C+E$ level for each parameter, as shown in the upper centre of Fig. 9.6a. This orders the parameters according to their interactive intensity. Finally the rock classification index can then be established as indicated at the bottom of Fig. 9.6a. The coefficients preceding each parameter are determined from the scaled $C+E$ values. This means that each parameter is weighted according to its interactive intensity.

The two cases illustrated in Fig. 9.6b illustrate how the number of parameters required will depend on the form of the parameter constellation in the *C vs. E* plot. The value of the parameter itself at a specific site can be assigned according to certain levels, as indicated via a 'pull-down' menu (see Fig. 9.6a). Thus, via the systems methodology, we have established a very logical, comprehensive and objective-based rock classification index that can be used for any rock engineering circumstances.

9.3 THE R^2 METHODOLOGY

At this stage, it is worth recapping on the overall systems methodology in order to review its components and to reflect on the developments so far in the book. In line with the acronyms already discussed, the total approach is termed REMIT/RESPONSE. The components of the methodology are listed immediately preceding Section 9.1 and in Fig. 9.7, with the acronym of the acronyms being termed R^2. It is important to refer back to the Icon at the start of this Chapter and to Fig. 9.1, and remember that this methodology is utilized in both the analytic and synthetic directions. The reader will recall that we enter a rock engineering design problem with ONSE, RESP and REMIT successively; we then exit with REMIT, RESP and ONSE successively.

REMIT, Rock Engineering Mechanisms Information Technology, is comprised of a knowledge of the basic parameters and mechanisms. This includes the leading diagonal terms and off-diagonal terms of the interaction matrix, forming the basic structure and connections of the system. It represents, therefore, the morphological and cascading types of system, i.e. the body and functional connections of the system.

RESP, Rock Engineering System Performance, is comprised of rock mechanics plus rock engineering. This refers to the actual dynamic operation of the matrix as a natural process-response system before engineering, and then also includes the perturbations introduced by engineering operations. RESP is the total operation of the matrix as a process-response system.

ONSE, Objective-based Network Sequence Evaluation, considers the project engineering itself. It is the control system superimposed on the natural process-response system. From the leading diagonal term, Construction, impulses are sent out in order to manipulate the matrix to achieve the target matrix, as defined by the objectives.

The operation of these three components is via the iconic representation at the beginning of this Chapter. It is essential that the methodology is implemented firstly analytically and then synthetically, as has already been described in this Chapter. Referring back to Icon 1 on page 1, we have to start analytically. We must define the system as containing everything. The key then is the interaction matrix device and the winnowing process illustrated in Fig. 9.5. Thus, the REMIT/ RESPONSE methodology is a sequence which has to be operated in the correct order to build up the optimal design and construction procedures. *It is essential to understand that a precursor to the operation of the methodology in Fig. 9.7 is the establishment of the project objectives.*

During the development of the rock engineering systems theory and practice described in this book, the author has been surprised to discover that the most difficult part of the whole process is establishing the objective. It may appear at

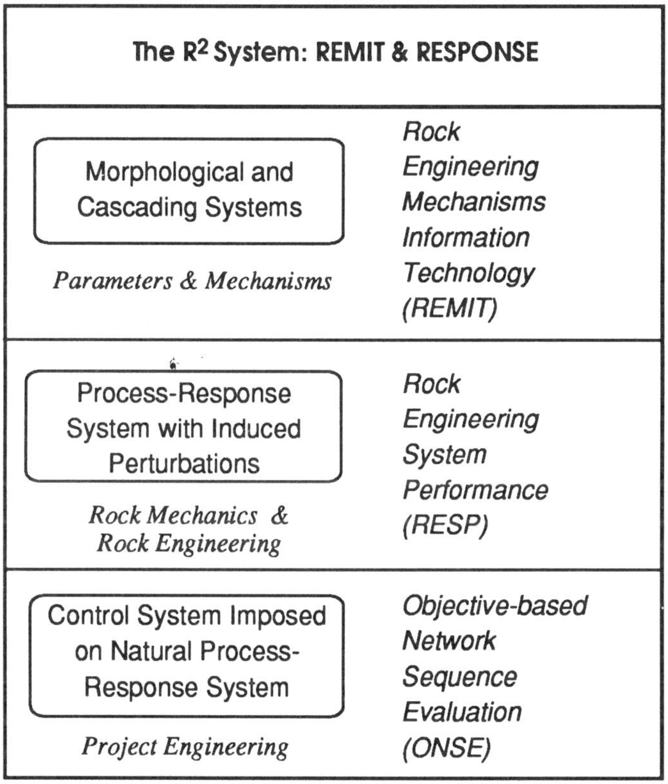

Fig. 9.7 The components of the R^2 methodology used to implement the systems approach to rock engineering.

first sight that the objective of a rock engineering project is simple. In general, this is not the case — because there is usually a hierarchy of objectives.

Consider the pre-split rock face illustrated in Fig. 9.8. This is an example of an excellent pre-splitting operation, which was conducted on the side of the A9 road near Perth in Scotland. The robustness of the technique is illustrated by the fact that the establishment of the pre-split plane has been completely successful, despite the presence of adverse discontinuities and even changing strata (note the ash flow in the lower part of the slope). Now, what was the objective of this pre-split plane? Clearly, it was to create an initial fracture plane so that a stable slope would be obtained following bulk blasting excavation of the main cutting. But why is a stable slope required? Is it for physical safety or is it to avoid chronic maintenance

Fig. 9.8 Example of a successful highway cutting pre-splitting operation, A9 road, Perth, Scotland.

requirements over the next few decades? The answer is both. Moreover, these two objectives are relatively compatible and so the engineering is straightforward. The successful design and management of the pre-split blasting has indeed led to the successful accomplishment of the objectives.

For comparison, however, imagine that the access tunnel illustrated in Fig. 9.9 were to be for an underground radioactive waste repository. What is the objective of the whole project? It is the successful isolation of radionuclides from the biosphere for a specified length of time. In the case of intermediate waste, the design life might be 500 years. What is the objective of an access tunnel to such a repository? It is to provide passage for the waste which is to be emplaced in the repository. The tunnel should therefore be stable during this phase of the repository's operation so that the emplacement can be successfully accomplished.

However, once the repository is full, the access tunnel has to be sealed in such a way that the total objective, isolation of the radionuclides, is maintained. At this stage, far from being of use to us, this tunnel now becomes a liability — because it is definitely a preferential high permeability route for the potential migration of groundwater. This tunnel has to be backfilled so that the permeability of the rock structure containing this tunnel is within the bounds dictated by the objective. Backfilling the tunnel is now driven by a totally different objective from that of the original construction.

Obviously, to ensure that both these objectives are successfully accomplished, we need to know what they are before we start on the construction process. This is why the establishment of the objectives and the ONSE component of the R^2 system are so critical. To some extent, although reconfigured, the REMIT and RESP components are the methods we traditionally use, formalized in this book as

Fig. 9.9 Access tunnel in the Alto Lindoso hydroelectric scheme in Portugal.

a coherent, structured and comprehensive methodology utilizing systems thinking. The winnowing process ensures that there is a manageable knowledge base and that it is dedicated to the project. However, the whole key is to conduct the engineering as conceptually illustrated in Icon 7 and Fig. 7.9, such that the target matrix is achieved.

More research needs to be conducted to establish a firm scientific foundation for coherently stating the objectives of a project. Furthermore, there are naturally interfaces with other disciplines: there are financial issues; there are environmental issues; there are social issues. If we state that the objective of mining is to provide the raw materials for civilisation, we can develop a hierarchy of objectives associated with this 'top of the flow chart' objective. If the objective of mining is to make the maximum profit, then we can do the same. The hierarchy of objectives for these two cases will be subtly, or even significantly, different.

The implications for rock engineering are enormous. We have arrived at a stage in describing the methodology where this single point, how to specify the objectives, has emerged as the most important problem now confronting us. This is reinforced as our own work becomes more integrated with other disciplines and is conducted within an increasingly global environment.

9.4 COMPUTERIZING THE METHODOLOGY

The basic interaction matrix device, the operation of the matrix, and the pathways through the matrix are eminently suitable for computerization. As this book is being written, in 1992, both the software and the hardware facilities available are increasing rapidly in their capabilities.

In Fig. 9.10, there is a fanciful listing of factors that would be involved in computerizing the whole methodology represented here. It is emphasised that these are a random selection of items that have arisen in the author's mind as the methodology has been generated and developed. Let us consider these in turn.

Line 10. A multimedia approach is required. The 'multimedia' term refers to the ability to have screen text, calculations, graphics, photographs, sound etc., all

Computerizing the Methodology

```
>LIST

 10 Multimedia
 20 Parameter Palette
 30 Transfer Function Algorithms
 40 Identity Icon
 50 Mechanism Editing
 60 N-tree Encoding
 70 Concatenation of Matrices
 80 Interactive Icons
 90 Coincidence of Thresholds
100 Compact Disc
```

Fig. 9.10 Some of the items that will have to be considered during full computerization of the implementational methodology.

as part of the computer package. This is a very tempting approach because the images in this book, and many, many more, could then be easily stored and recalled at the click of a mouse. The difficulty in such multimedia computerization is the amount of work that would be required to establish the multimedia facility.

Line 20. Similar to the palettes in word processing and drawing programmes, the R^2 program could have a parameter palette allowing choice of the leading diagonal terms in the matrix. Alternatively, all parameters could be selected and then each considered for deletion via the winnowing process. Many of the operations described in the preceding Chapters could be undertaken in the same way that word processing programs operate.

Line 30. A suite of transfer function algorithms could be included to specify the mathematics and mechanics of off-diagonal boxes. These could be selected from a mechanism palette.

Line 40. Can a parameter interact with itself? Can the discontinuity aperture affect the discontinuity aperture? Have we established that the leading diagonal terms should be coded as zero? What is the identity icon? Is it possible to have an interaction matrix with all off-diagonal terms being zero — for certain 'eigenvalues' of the leading diagonal parameters P_i ?

Line 50. As the matrix travels through time and thresholds may be reached, mechanisms can be edited. Perhaps several mechanisms can be incorporated in each off-diagonal box, depending on the resolution of the matrix.

Line 60. The resolution of the matrix could be automatically varied, either hierarchically expanded or contracted via N-tree encoding. Some form of automated matrix resolution is required, similar conceptually to automatic finite element mesh generation.

Line 70. We need to understand the development of the matrices through time. The matrices could be incrementally concatenated and output to correspond to various stages in the project. Conversely, proformas could be generated to provide input at various stages of concatenation in order to calibrate the matrix and to ensure that there is a constant feedback control mechanism.

Line 80. Iconic interface programmes are now appearing in the market place, ideally suited to the linkages of the terms within the matrix. Moreover, these would lead to very user-friendly methods of operating programs for the systems analysis and synthesis.

Line 90. Accidents usually occur because certain combinations of circumstances are not expected to arise. We ought to be able to anticipate anything and everything with the systems approach. The right hand side of Icon 1 represents a model that, by definition, contains everything. Any coincidence of thresholds that might be classed as 'unexpected' can be established by 'letting the matrix run', whether deterministically or probabilistically.

Line 100. Finally, with the advent of compact discs, we have as much storage space as we could possibly want. All the existing rock mechanics books could be stored on one compact disc. We could store all abstracts of all the papers that have ever been published in rock mechanics on one compact disc. We can store a large number of photographic images on one compact disc.

The very nature of computer software is changing. Neural networks are being simulated so that the computer can mimic the interpretative capabilities of the human brain, with similar learning and pattern recognition abilities. How does the human brain recognise a specific human face from all the faces that it knows in a split second? How does the human brain recognise an abstract picture of a train, when no features of the train are present except the 'essence' of the train? What is this 'essence'? How does the sum become greater than the parts? If the network analysis presented earlier is linear, i.e. two mechanistic pathways can be

The Link with Neural Networks

Detail of Neuronal Architecture

S_i - incoming signals

w_{ij} - weighting of pathway (synaptic efficacy)

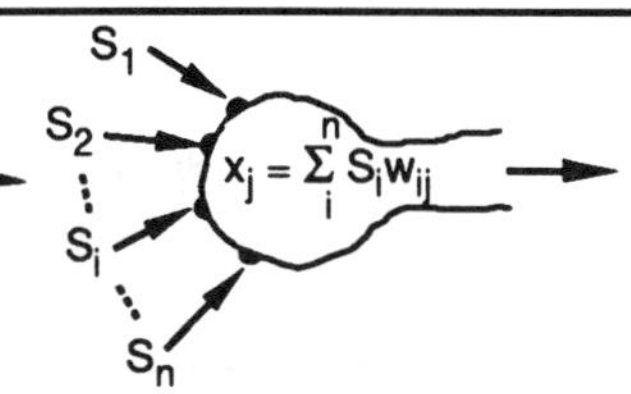

Neural networks are trained with examples of concepts to capture

There are seven basic aspects of neural networks:

1. A set of processing units
2. The state of activation of a processing unit
3. Function used to compute output of a processing unit
4. The pattern of connectivity among processing units
5. The rule of propagation employed
6. The activation function used
7. The rules of learning

Many aspects of neural networks can be modelled via the matrix R^2 system

Fig. 9.11 The interaction matrix and the associated pathways through the matrix are strongly related to neural networks.

superimposed, how do we achieve a holistic phenomenon? Are rock engineering accidents caused by holistic phenomena, where the sum of two mechanisms is, in fact, greater than the two mechanisms themselves? (For further information on neural networks, the reader is referred to "An Introduction to Neural Computing" by I. Aleksander and H. Morton, Chapman & Hall, 240pp., 1990.) The main basic aspects of neural networks are illustrated in Fig. 9.11. All these can be modelled via the matrix R^2 system and its extensions. It is highly likely that we will need neural networks and associated sophisticated programming capabilities in the computerization of the rock engineering systems methodology, which has not yet been attempted on an overall strategic level.

10

Systems Thinking

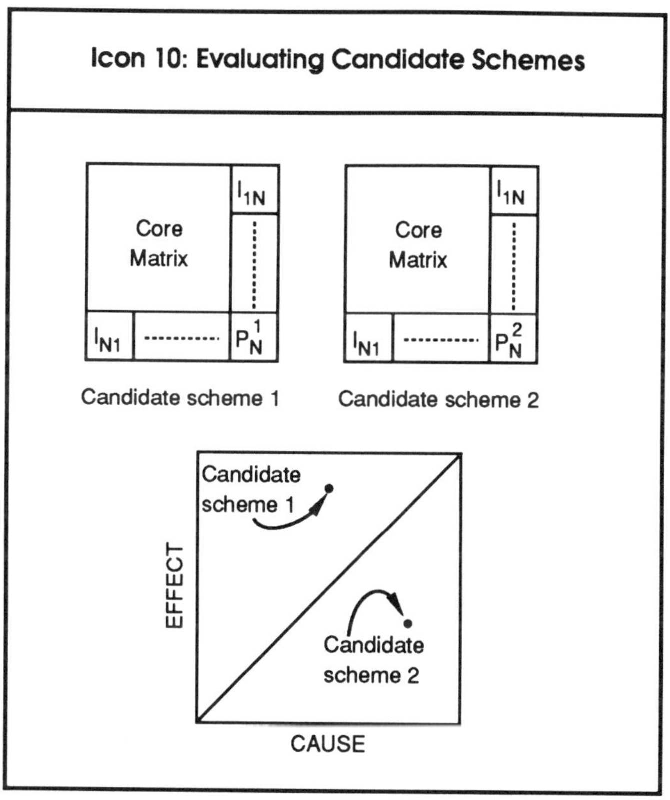

From the rock mechanics core matrix and the features of a particular engineering proposal, we can establish whether the project will control the rock mass or whether the rock mass will control the project.

In the previous nine chapters, the concept of rock engineering as a system has been described. Although many of the devices that have been presented in the book so far are novel, systems concepts have been widely applied in a variety of subject areas. In many ways, engineering is a much less complex subject than some of those with which mankind has wrestled in the past. Questions associated with the nature of life, moral issues and social systems are far more complicated.

As rock engineers, we only have to understand the Rock Engineering System (RES) as a natural process–response system, and then to consider how the system can be appropriately perturbed to achieve the project objectives. We do not have to be concerned why the earth is here in the first place, nor indeed why the geological structures are exactly the way they are. However, in the fullness of time, it may well enhance the implementation of rock engineering systems methodology if we do know more about the history of the earth and are able to link rock engineering projects with the wider issues of environmental, social and even planetary benefit.

However, we should not be too optimistic just yet but heed some of the words in a book written in 1699 by Dr Thomas Burnet entitled "The Theory of the Earth". In the Introduction to his book, he states "... all I say, betwixt the first Chaos and the last Completion of Time and all Things temporary, this was given to the disquisitions of Men: on either hand is Eternity, before the world and after, which is without our Reach: but that little Spot of Ground that lies betwixt those two great Oceans, this we are to cultivate, this we are masters of, herein we are to exercise our Thoughts, to understand and lay open the Treasures of the Divine Wisdom and Goodness hid in this Part of Nature and of Providence."

10.1 ENGINEERING CONTROL

So, with the previous chapters behind us and with Dr Burnet's motivation, let us consider how systems thinking can assist us on site at our little Spot of Ground. All rock masses that are going to be subject to future engineering, whether controlled by us or future generations, are already in existence. As has been described, the matrix representing the natural system can be compiled. So, when activating the Construction box in the matrix, how do we perturb the rock to achieve the target matrix?

As shown in Fig. 10.1, there are four possibilities in the choice of engineering control techniques, i.e. the superimposition of an intelligent control system on the natural process–response system. Remember that a parameter can have a low interaction intensity or a high interaction intensity. If the system interaction intensity is low, as indicated by the top left cause *vs.* effect plot in Fig. 10.1, the parameters have a large degree of independence. Conversely, at the top right of Fig. 10.1 in the cause *vs.* effect plot, a system with a high interaction intensity is represented. Most parameters will be highly dependent on the system performance because of the interaction within it. This immediately indicates the division of engineering control techniques into 'direct' and 'indirect' methods. If a parameter is largely independent, then engineering control can be applied directly to the parameter — if physically possible. On the other hand, if a highly interactive parameter is being

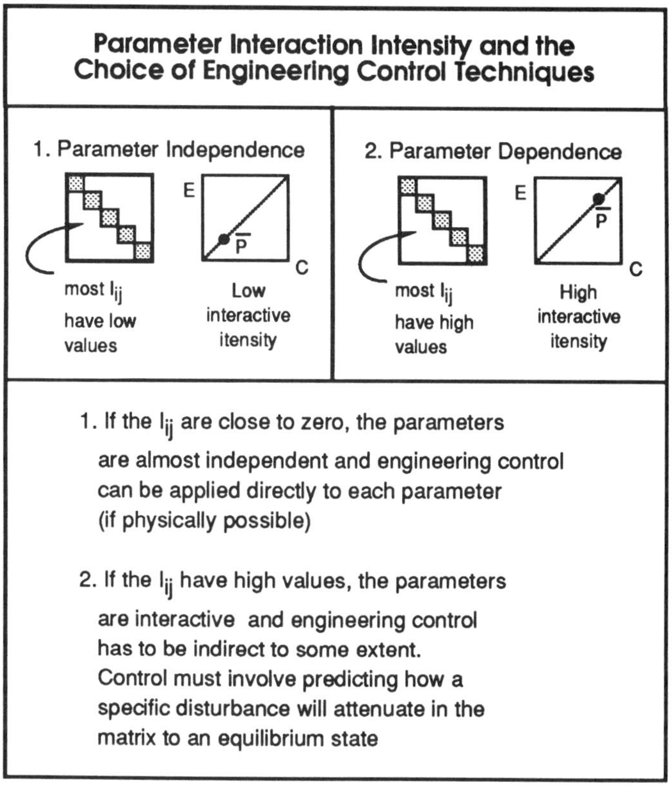

Fig. 10.1 The choice of engineering control technique must depend on the interaction intensity of the parameters.

considered, the engineering control has to be indirect to some extent. If any attempt is made to apply control directly to the parameter, this will cause a perturbation to the whole matrix, which could counteract the attempted control.

By consideration of the rock engineering system and its representation via the interaction matrix, we conclude that there are four main engineering choices:

1. alteration of one or more parameters;
2. alteration of one or more interactive mechanisms;
3. leave the parameters and mechanisms as they are and achieve the target matrix by the type and sequence of engineering construction; or
4. any combination of the above.

Whichever of these we choose, the construction box must be activated via one of the five basic forms of input represented in Fig. 10.2. An example of a transient impulse is blasting; an example of a low level input energy rate is a tunnel boring machine.

A complete description and systems analysis of these four engineering choices is beyond the scope of this book so we will illustrate the ideas by highlighting parameter alteration techniques.

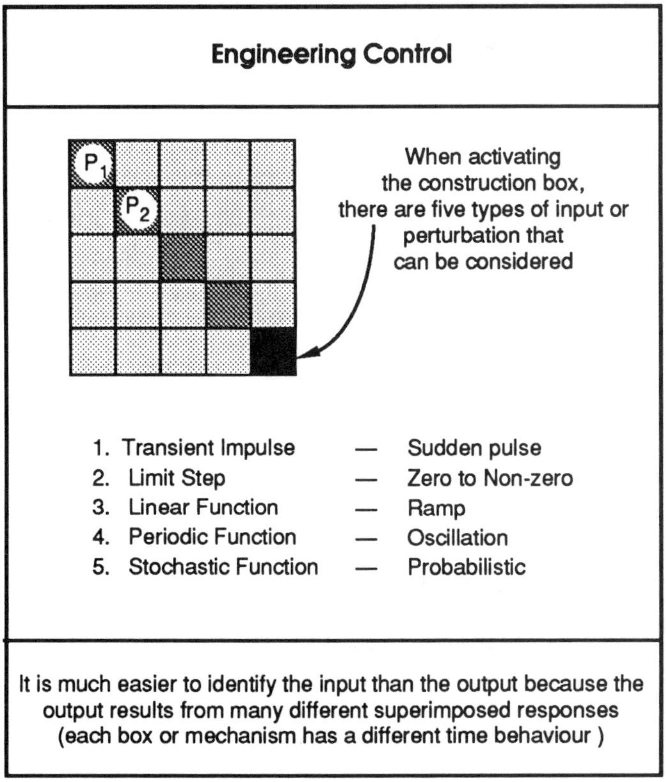

Fig. 10.2 Once the overall engineering control method has been established, the actual control can have one or more of the forms above.

10.2 PARAMETER ALTERATION TECHNIQUES

The first of the engineering control techniques, alteration of a parameter, arises as a consequence of the Rock Engineering System Performance (RESP) and Objective-based Network Sequence Evaluation (ONSE) considerations. An extreme case of parameter alteration is removal of a particular parameter as represented in Fig. 10.3. This has a dramatic effect on the behaviour of the matrix (assuming that the parameter is a significantly interactive one) because the process results in deletion of all mechanisms in the row and column passing through the leading diagonal parameter in question. Naturally, the removal of these terms will potentially affect all the cause and effect values for all the other parameters.

The effect of the removal of a parameter from the *C,E* diagram is shown in Fig. 10.4. If the parameter in question is removed entirely, the incremental changes to each of the other parameters can be represented by the small vectors in Fig. 10.4. Naturally the mean parameter value will also reduce (if the original matrix dimension is retained) but, from Theorem 1, will maintain its position as the centre of gravity for the remaining points.

The reader may well ask how it is possible to delete, or indeed even alter, any parameter. In Fig. 10.5, the twelve parameters used in the generic slopes matrix in Chapter 3 are reproduced — with an additional column stating whether these can be altered in any sensible engineering way. It transpires that six of the parameters are essentially fixed, whilst the other six can be altered as part of the engineering. The distinction is not clear cut: some of these parameters are less fixed than others. Referring to the first parameter, the Overall Environment, we cannot significantly alter the geology, climate, seismicity or previous instability. Referring to the last parameter, Construction, we can alter the method and sequencing of the excavation technique.

The pre-existing discontinuity geometry cannot be altered although, as we shall see, it is possible to alter the mechanical properties of the discontinuities. In this list, the slope orientation and location and dimensions have been listed as essentially

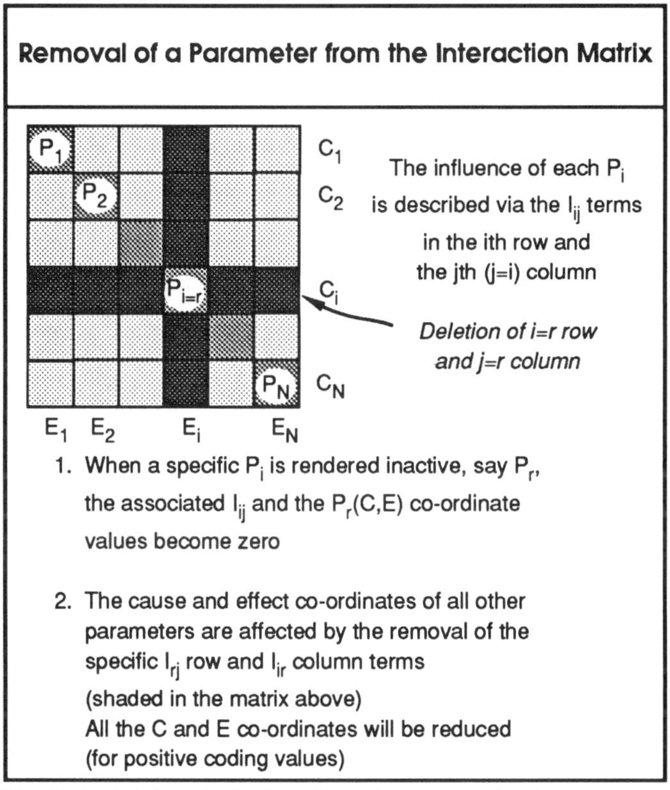

Fig. 10.3 Deletion of a leading diagonal parameter also deletes the row and column through the parameter — eliminating mechanisms.

fixed. Obviously this will depend on the type of engineering. In a road cutting, there are usually strong constraints, but in a mine there is usually more flexibility.

The engineering removal of an interactive parameter from the generic slope stability matrix (Section 3.1) is illustrated by the example of totally removing the fourth parameter on the leading diagonal, the mechanical properties of the discontinuities — as illustrated in Fig. 10.6. In practice and in suitable circumstances, this can be achieved by washing out the discontinuity fill and then grouting the discontinuities. This eliminates 22 interactive terms in the 12x12 matrix and provides a barrier for many mechanism pathways. These 22 interactive mechanisms and the benefit of their removal are listed in Fig. 10.7.

This specific operation was undertaken at the Fei-Tsui Dam in Taiwan, shown in Fig. 10.8. It is a wide concrete dam, 122.5 m in height and 510 m crest length.

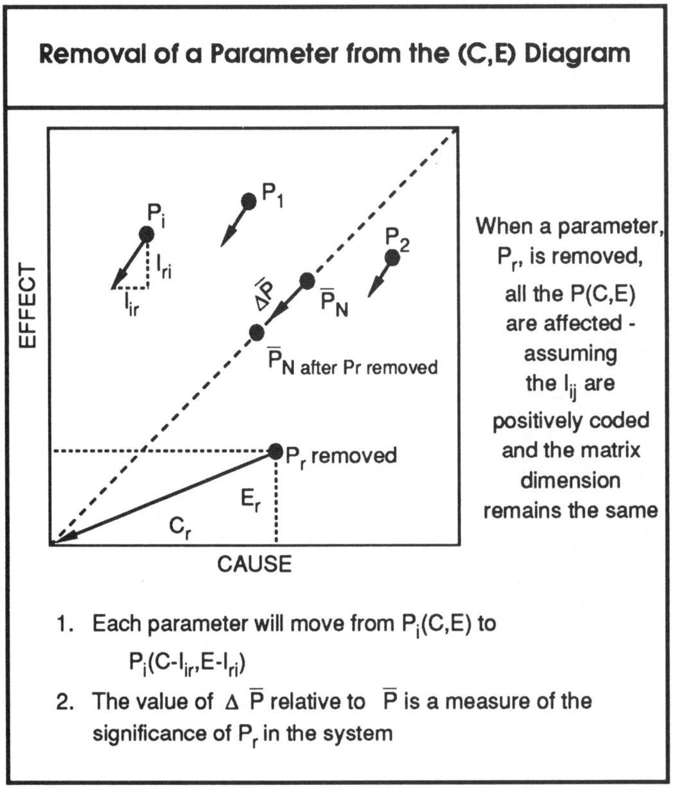

Fig. 10.4 Deletion of a parameter from the matrix influences the cause and effect co-ordinates of all the other parameters.

Massive sandstone and siltstone of Oligocene age form the foundation of the dam site. The strike of bedding is generally parallel to the river with a dip of 40° to the right bank. This influences the left abutment dip slope and the right scarp slope. Although the sandstone and siltstone are considered to be strong, they contain clay bedding seams varying from less than 10 mm to about 150 mm in thickness and form unfavourable weak planes.

Since the dam is located only 30 km upstream of the densely populated city of Taipei, dam stability is especially important. The rock mass shear strength at the dam site is the key parameter. In practice, there was little that the engineers could do to improve the quality of the intact rock, and that left the engineering effort directed towards the improvement of the mechanical properties of the discontinuities.

These were effectively removed from the problem, exactly as illustrated in Fig. 10.6 and described in Fig. 10.7, by use of a water jet and water knife to wash out the discontinuity fill, which was mainly clay, rock fragments and weathered rock. When the aperture of the discontinuity was larger than 10 mm, a water jet was used between two parallel adits approximately 12 m apart. When the aperture of the discontinuity was smaller than 10 mm, a water knife and water drill were used, with a water drill hole spacing at about 200 mm. The water jet operated at a pressure of 20 MPa; the water knife and water drill operated at a pressure of 240 MPa.

Possibilities for Altering the 12 Parameters in the Generic Slopes Interaction Matrix (Fig. 3.1a)

	Parameter	*Ability to alter*
P1	**Overall Environment** Geology, climate, seismic risk, previous instability	Cannot be altered
P2	**Intact Rock Quality** Strong, weak, weathering susceptibility	Cannot be altered
P3	**Discontinuity Geometry** Sets, orientations, apertures, roughness	Cannot be altered
P4	**Discontinuity Mechanical Properties** Stiffness, cohesion, friction	Can be altered
P5	**Rock Mass Properties** Deformability, strength, failure	Can be altered
P6	***In Situ* Rock Stress** Principal stress magnitudes and directions	Cannot be altered
P7	**Hydraulic Conditions** Permeability	Can be altered
P8	**Slope Orientation and Locations** Dip, dip direction, position	Essentially fixed
P9	**Slope Dimensions** Height, width, (local and overall)	Essentially fixed
P10	**Proximate Engineering Disturbances** Adjacent blasting	Can be altered
P11	**Support/Maintenance** Bolts, cables, grouting, pre/post construction	Can be altered
P12	**Construction** Excavation method, sequencing	Can be altered

Four of the parameters cannot be altered, two might be specified, and six can be altered by or as part of the engineering

Fig. 10.5 Consideration of the extent to which the leading diagonal parameters in the generic slopes matrix (Section 3.1) are fixed.

Naturally, this discontinuity treatment programme initially required excavation of a series of access tunnels and working adits in the zone loaded by the dam (foundation and abutments). A typical discontinuity is shown in Fig. 10.9a; a 'half-way stage' in the washing out is shown in Fig. 10.9b; and a cleaned, open discontinuity is shown in Fig. 10.9c. Note the cap lamp light visible through the cleaned discontinuity.

This remedial programme was extensively planned to allow for the idiosyncrasies of the left and right abutments and the zone below the river bed. The washed and cleaned discontinuities were then backfilled with non-shrinking cement and grouted under pressure, with checks to ensure that the discontinuities had been completely filled. Tests on the shear strength showed that the treated discontinuities had good average Mohr-Coulomb strength parameters: a cohesion value of above 3.1 MPa

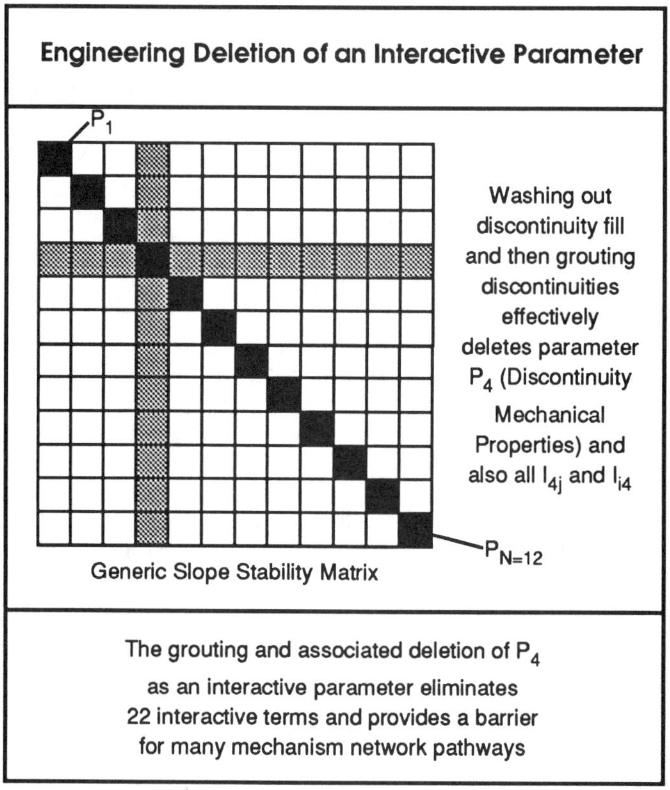

Fig. 10.6 Deletion of Parameter 4 (Discontinuity Mechanical Properties) in the generic slopes interaction matrix.

and a friction angle of 38°. The grouting access tunnels and boreholes were also backfilled with concrete and mortar. All this proved to be most effective and I am most grateful to Dr J Sheng for the information and to the RET-SER Engineering Agency and Dr C. T. Lee for the providing the photographs in Figs. 10.8 and 10.9.

The effective removal of the discontinuities in the case example just described is one of a large number of possibilities that are open to us. It is through consideration of the matrix and the options described in Fig. 10.1 that our imagination can be assisted and we can develop elegant and indeed optimal approaches to any rock engineering project.

This is one of the main advantages of systems thinking; the framework provides the structural basis for creative thinking. Whether the resultant design process is a science or an art is a moot point. Suffice it to say that we need to understand

Interactive Mechanisms Removed by Deletion of Parameter 4 in Slopes Matrix

Deletion of P_4 (Discontinuity Mechanical Properties)

as a result of washing out infill and then grouting removes or minimizes all I_{4j} and I_{i4} as follows:

Matrix *row* 4 interactions	Matrix *column* 4 interactions
4,1 Treated discontinuities lead to better overall environment	1,4 The overall environment has less effect on the discontinuities
4,2 Joint movement inhibited, less effect on intact rock	2,4 Weak intact rock has less effect on discontinuities
4,3 Strong joints inhibit formation of new joints	3,4 Number of effective discontinuities reduced
4,4 Discontinuity mechanical properties— enhanced by treatment described	*4,4 Discontinuity mechanical properties — enhanced*
4,5 Stiffer discontinuities give stiffer rock mass properties	5,4 Discontinuities less affected by mass behaviour
4,6 Treated discontinuities have less effect on the *in situ* stress	6,4 Discontinuities less affected by *in situ* stress
4,7 Filled discontinuities inhibit water flow	7,4 Discontinuities less affected by water flow
4,8 Stronger discontinuities give safer slopes	8,4 Position of slope will not affect discontinuities
4,9 Increased cohesion allows higher slopes	9,4 Discontinuities not susceptible to high slope effects
4,10 Stronger joints allow closer proximate disturbance	10,4 Discontinuities less affected by blasting
4,11 Treated joints require less attention	11,4 Less support required
4,12 Stronger joints allow greater flexibility during construction	12,4 Many construction problems minimized

Fig. 10.7 The 'interactive mechanisms' result of removing Parameter 4 from the generic slopes matrix.

the system in order to understand the engineering, in order to generate these 'lateral thinking' ideas, in order to confirm that the consequences of our engineering design are in line with our objectives, and that we will successfully reach the target state.

10.3 ASSESSING CANDIDATE SCHEMES

Given a rock engineering project and two candidate schemes submitted, how do we decide which is the best? One of the ways to approach this problem via the systems overview is to consider the actual *C,E* co-ordinates of the project itself, as shown in Icon 10 at the beginning of this Chapter and in Fig. 10.10. There is the natural process–response rock mechanics core matrix. We activate the project, by convention the *N*th leading diagonal term, and calculate its *C,E* co-ordinates in the usual way.

Fig. 10.8 The completed Fei-Tsui dam in Taiwan.

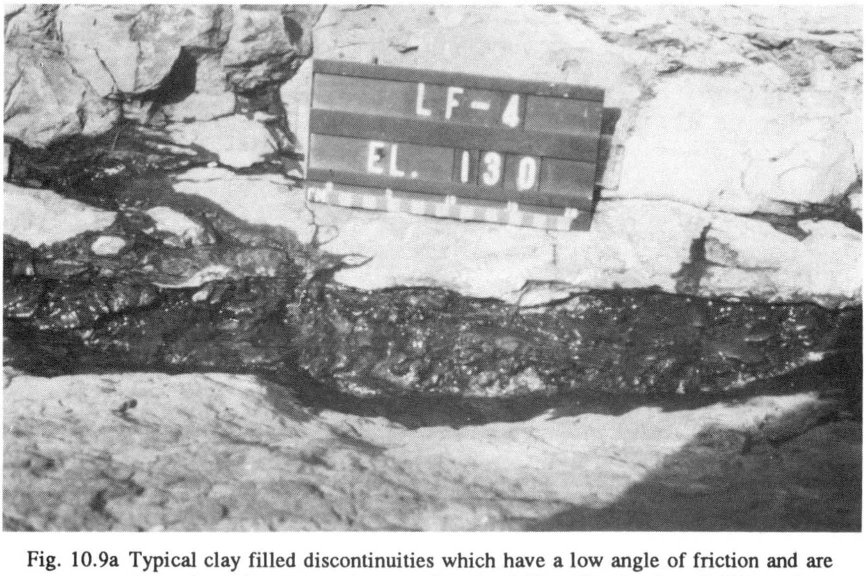

Fig. 10.9a Typical clay filled discontinuities which have a low angle of friction and are the main rock mass weaknesses (length of identification board: 200 mm).

Fig. 10.9b Washing out the clay in the discontinuities, by water knife for narrower apertures and water jet for wider apertures (length of identification board: 200 mm).

The project can then be plotted on the cause *vs.* effect diagram along with all the other parameters. How interactive is the project? Does it dominate the other parameters or do the other parameters dominate it? This is further clarified in Fig. 10.11 in which we are considering whether, through two binary interactions, the rock mass is dominating the project or the project is dominating the rock mass.

Two candidate schemes are shown in Fig. 10.12 with the cause and effect co-ordinates of the candidate schemes — indicating directly which of the candidate

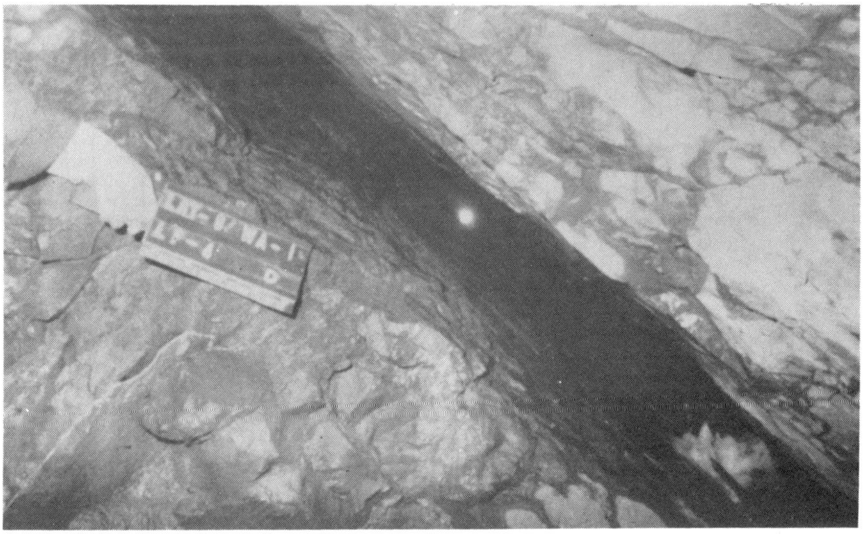

Fig. 10.9c View through the 'cleaned out' discontinuity prior to backfilling with cement mortar grouted under pressure (Length of identification board: 200 mm).

schemes dominates the rock mass more than the others. It does not necessarily follow that this is the best scheme but it certainly demonstrates strategic systems thinking. Naturally, this is the first step in a systems evaluation of the schemes which will include analysis of matrix pathways and whether the implementation of a scheme could lead to unacceptable combinations of parameters linked via a pathway.

Indeed, it could well happen that the consequence of this type of analysis determines the research and/or the site investigation necessary prior to choosing the scheme — because it may not be entirely clear how well the mechanisms in the *N*th

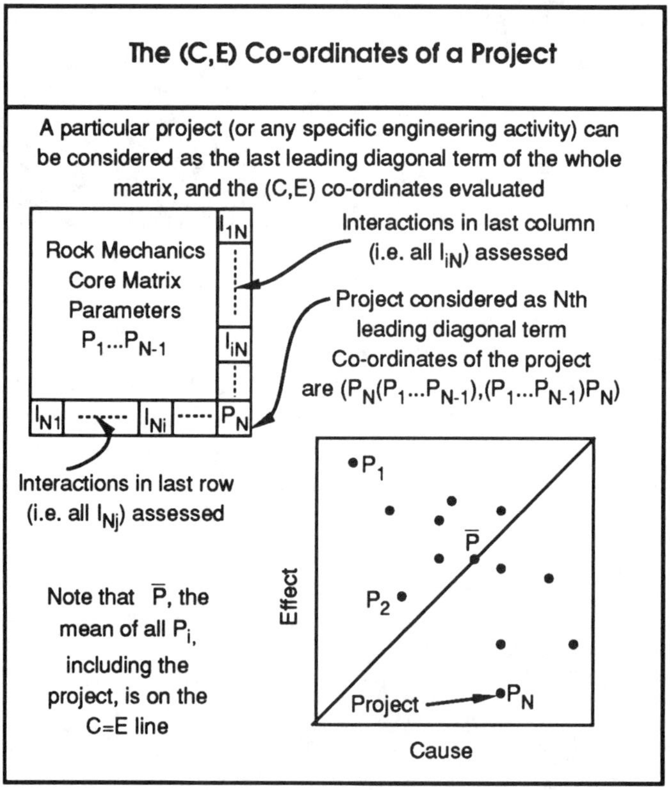

Fig. 10.10 A project has *C,E* co-ordinates and can be plotted on the cause *vs.* effect diagram with the leading diagonal parameters.

row and column are understood, nor the consequences of the matrix mechanism pathways.

It is necessary to have sufficient knowledge for such analysis but for some non-precedent practice projects this knowledge may simply not be available. The purpose of the Underground Research Laboratory operated by the Atomic Energy of Canada Ltd. in Pinawa, Manitoba is to provide a site facility for experiments to address the technical information required for radioactive waste disposal. The Laboratory itself will not become a repository: the work there is focussed on understanding *in situ* rock stress, rock structure and the effect of these on rock mass permeability. The main technical issue in radioactive waste disposal is the prevention of unacceptable levels of radionuclide migration in groundwater, which has to be modelled adequately in order to be able to assure the long term safety of a repository. Naturally, the excavation process should not significantly affect the near field permeability and an example of high quality smooth wall blasting at this site is shown in Fig. 10.13.

The disposal of radioactive waste is a problem currently also being adddressed in the UK, which has nuclear power stations. The Trawsfynydd power station in Wales is shown in Fig. 10.14. For this particular application, and indeed all rock

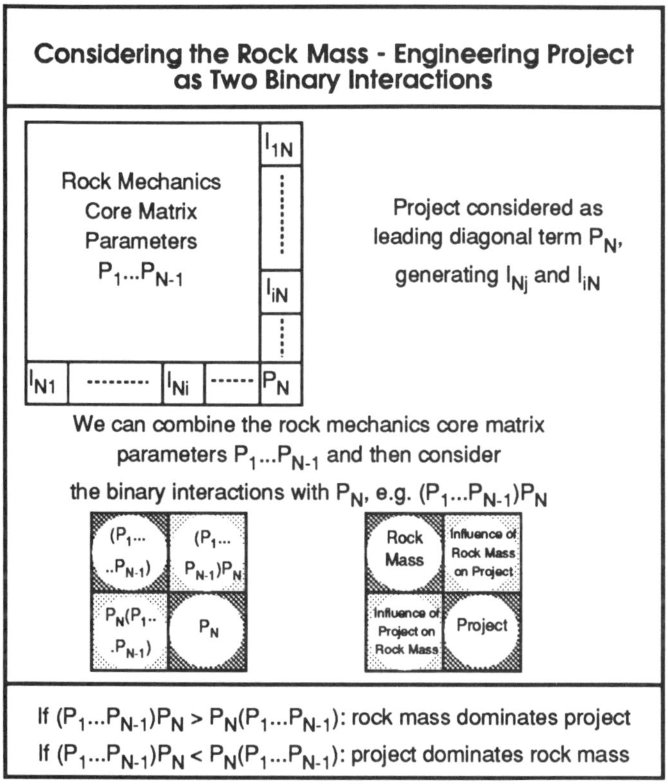

Fig. 10.11 Does the project dominate the rock mass or does the rock mass dominate the project?

engineering projects, the strategic assessment of project robustness should be established. As a systems overview, the implications of the interaction intensity and dominance of each proposed scheme should be considered as illustrated in Fig. 10.15 — further extrapolating the ideas in Fig. 10.12. Such an assessment would be conducted early on as an initial assessment of viability. After establishing which schemes appeared acceptable, the full force of the methodology would be applied to them. This whole subject leads us naturally into the concept of systems auditing.

10.4 SYSTEMS AUDITS OF ROCK ENGINEERING PROJECTS

From all the information presented in this book, it is clear that an engineering project could be implemented such that it leads to disastrous consequences or, by using other techniques, the work could be completed optimally — with the least money, greatest benefit to the environment, using the least energy, and creating the least disorder. How can engineering projects be audited to establish their 'efficiency', before, during, and after construction? Most projects are now designed and constructed following quality assurance procedures. However, these only ensure that the quality of the work is assured once the procedures have been established. A more general audit of the whole system is required. There is no point in quality assuring ourselves into a disaster. As projects become larger, this subject becomes increasingly important, because the projects become more difficult to control. (The

Fig. 10.13 High quality smooth wall blasting on the 240 m level of the Underground Research Laboratory, Pinawa, Manitoba, Canada.

reader is referred to the discussions in "Macro-Engineering: Global Infrastructure Solutions", Eds. F. P. Davidson and C.L. Meador, Ellis Horwood, 194pp., 1992.)

Assuming that the fundamental rock mechanics has been satisfactorily implemented via the R^2 methodology, the four main possibilities for rock engineering audits are listed in Fig. 10.16. The most well known audit, and the easiest to conduct is the financial audit. The matrix representing the system is in a certain state and it is changed by man to another state. This could be for a small increment in time or for the whole duration of the project from beginning to end. Generally, it is possible to discover the financial cost of the project components — although, as all practitioners know, there is a huge variety of qualifiers to this statement. Moreover, the financial audit should really be a resource audit. It is not only the cash that is required but the items that are being purchased. What are the resource implications

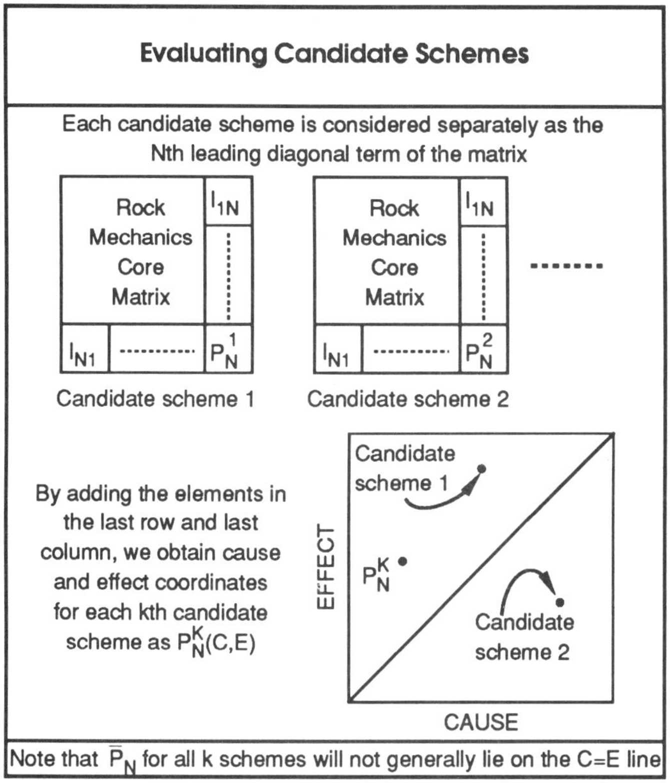

Fig. 10.12 Comparison of k competing candidate schemes (1 & 2 highlighted) based on their position in the cause *vs.* effect diagram.

Fig. 10.14 The only inland nuclear electricity generating power station in the UK, at Trawsfynydd in Wales.

of using a particular item? The financial practicalities and the wider issues need to be audited, both in concept and as operated during construction.

The second audit, the environmental audit, again should, *in theory*, be easy to conduct. In fact, if the matrix does correctly represent the total system and its components are understood, then by definition we ought to know all the physical changes that will occur as a result of the project engineering. Of course, it is necessary to understand the operation of the matrix in order to assess all the factors, which becomes progressively more difficult as the matrix behaviour is extrapolated

Strategic Assessment of Project Robustness (both absolutely and as a comparison of schemes)

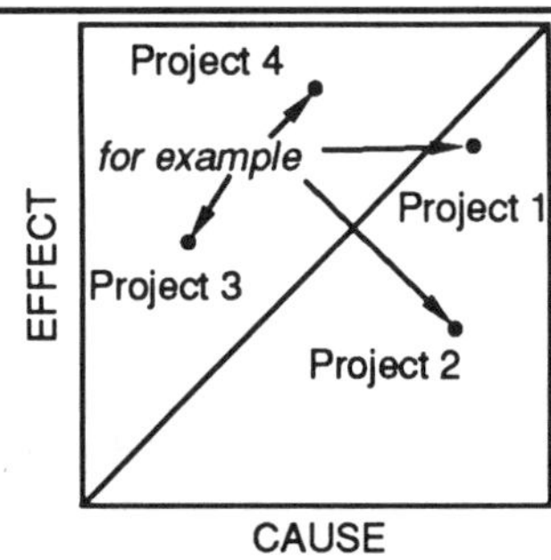

Each project is plotted via its cause and effect co-ordinates $P^K(C,E)$

The robustness of the project can be assessed strategically from its position in Cause–Effect space

Project 1: Strong interaction with the rock mass; project slightly dominates rock mass

Project 2: Medium interaction with the rock mass; project strongly dominates rock mass

Project 3: Weak interaction with the rock mass; project slightly subordinate to rock mass

Project 4: Medium to strong interaction with rock mass; project very subordinate to rock mass

Note: $P^K(C,E) = P^K((P_1 \ldots P_{N-1})P_N, P_N(P_1 \ldots P_{N-1}))$

Fig. 10.15 Four project positions on the cause *vs*. effect diagram illustrating rock mass–project interaction and dominance possibilities.

further forwards in time. However, the environmental audit is very important because of the local, regional, national and international consequences of the project.

The third type of audit listed in Fig. 10.16 is the energetic audit. How much energy is used in the project? This is not always easy to establish. For example, we know that less than 1% of the energy used by tunnel boring machines is actually used to break the rock. The rest is lost in various ways: stress waves in the rock, friction, heat, etc. However, all this can be simply short-circuited by considering the energy that has been supplied. This does not apply to items brought on site and it can become difficult to back track on the energy used for example in making a pre-cast tunnel lining segment. If the availability of energy is not an issue, then perhaps this audit is less important — in the rock engineering project context, but not in the wider context. A responsible engineer should consider whether there are any major energy resource implications of a particular rock engineering project.

Finally, we arrive at the entropic audit which has already been introduced in Chapter 8. This audit considers how much energy has been converted from a usable form to a non-usable form, and what degree of disorder has been introduced by the engineering. We noted earlier that the process of engineering will, within the total system context, always create more disorder than order — because of the

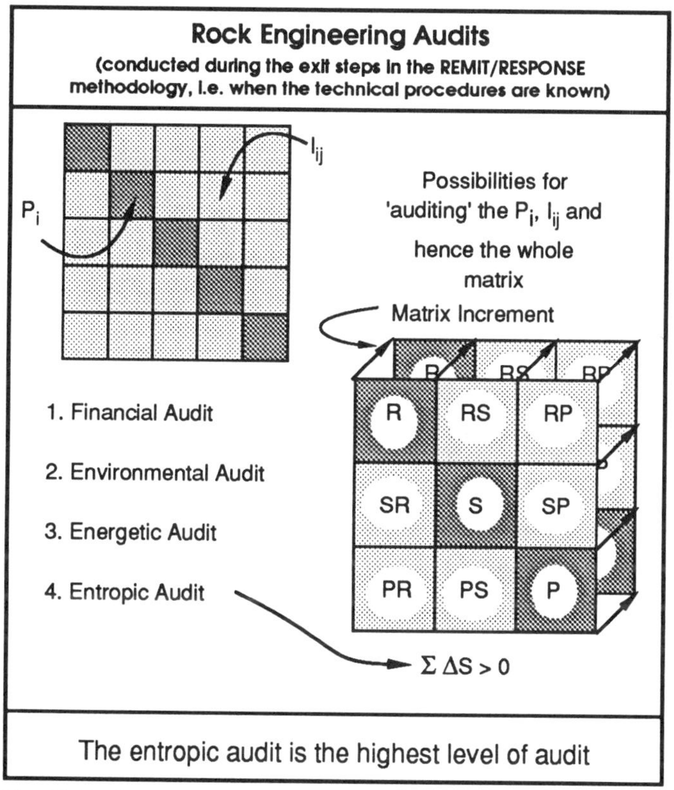

Fig. 10.16 The systems approach to rock engineering automatically suggests the idea of 'systems auditing' of projects.

monotonic increase in entropy associated with all mechanisms. It is easy to identify the order created by the engineering. The difficulty may be to identify the more than compensatory disorder that has been induced both in the proximity of the engineering works and further away. The entropic audit is the highest level of audit: it is both very subtle and very important.

In the entropy context, the history of the second law of thermodynamics and its relation with biology and man's activities has been fraught with discussion, argument and inconclusiveness. Everything ought to gradually disintegrate and become uniform. This has been recognized by many people including Dr Burnet in his 1699 book on the theory of the Earth which has already been quoted. Why is life becoming more complex? Compounds are created and combined forming life of exquisite complexity and function. In biology, the entropic ramifications of life have been identified, i.e. the disorder associated with life's order. But, so far, this has not yet been comprehensively attempted with engineering, the environment, or

Fig. 10.17 The surface of Mars viewed from Viking Lander 2 on a sunny afternoon at -150°F.

national planning. All these would benefit greatly from entropic audit analysis. From the strategic planning perspective, there is no point in completing the rock engineering project at all if the disbenefit associated with the disorder is greater than the benefit associated with the order. Thus, systems auditing is going to become increasingly important as the impact of engineering schemes (whether financial, environmental, energetic or entropic) crosses national boundaries and has to be considered in a global context.

In Fig. 10.17 there is a photograph showing carbon dioxide frost on boulders on the surface of Mars taken by Viking Lander 2 on a sunny winter afternoon when the temperature was -150°F. Does basic mechanics as we understand it operate on Mars? The answer will probably be 'yes' but the emphasis on the mechanisms will be different. The systems approach will still apply. The matrix approach will still apply. Indeed everything in this book will apply — except that the matrix may be rather sparse in terms of our knowledge! We know very little about the behaviour of rocks under conditions of extreme heat and cold.

Similarly, not only may we extrapolate in space but also in time. What will be the end-product of all our engineering? Everything that man makes will decay. Entropy will always have the upper hand. Engineers mine for minerals and obtain

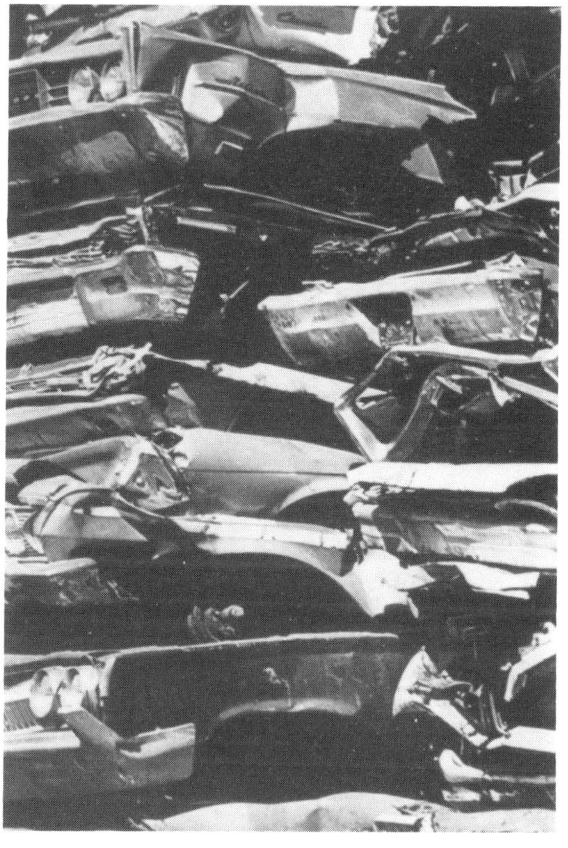

Fig. 10.18 Transient use of the earth's resources - an inevitable result of the entropic principle?

the metals that support civilization. The metals are an inherent and essential support for the function of, for example, the automobile industry; but a car's life is only an infinitesimal fraction of the life span of the metal itself. So where is the resting place of these metals after their brief interlude racing around the highways of the world, or perhaps stationary for much of the time in traffic jams? Some of them are piled up in the scrapyard shown in Fig. 10.18. Maybe the newer ideas of recycling and substitution should be introduced into our systems analysis —and, once again, we arrive back at the complications of establishing the objectives of the project.

It was mentioned earlier in this book that the single most unexpected aspect of the theory described here is the difficulty of establishing the overall objective, together with the hierarchy of objectives. The photograph in Fig. 10.18 graphically illustrates this problem. What exactly does the systems analysis say about an engineering project which, in the fullness of time, simply moves metals from one part of the earth to another? In the process, a great deal of money is involved, the environment can be severely disrupted, and entropy has been increased. This is perhaps a question for the philosophers and politicians; the author is going to concentrate on implemention of the systems approach as the next research phase.

Index